Communications in Computer and Information Science 2947

Series Editors

Gang Li, *School of Information Technology, Deakin University, Burwood, VIC, Australia*

Joaquim Filipe, *Polytechnic Institute of Setúbal, Setúbal, Portugal*

Zhiwei Xu, *Chinese Academy of Sciences, Beijing, China*

Rationale

The CCIS series is devoted to the publication of proceedings of computer science conferences. Its aim is to efficiently disseminate original research results in informatics in printed and electronic form. While the focus is on publication of peer-reviewed full papers presenting mature work, inclusion of reviewed short papers reporting on work in progress is welcome, too. Besides globally relevant meetings with internationally representative program committees guaranteeing a strict peer-reviewing and paper selection process, conferences run by societies or of high regional or national relevance are also considered for publication.

Topics

The topical scope of CCIS spans the entire spectrum of informatics ranging from foundational topics in the theory of computing to information and communications science and technology and a broad variety of interdisciplinary application fields.

Information for Volume Editors and Authors

Publication in CCIS is free of charge. No royalties are paid, however, we offer registered conference participants temporary free access to the online version of the conference proceedings on SpringerLink (http://link.springer.com) by means of an http referrer from the conference website and/or a number of complimentary printed copies, as specified in the official acceptance email of the event.

CCIS proceedings can be published in time for distribution at conferences or as post-proceedings, and delivered in the form of printed books and/or electronically as USBs and/or e-content licenses for accessing proceedings at SpringerLink. Furthermore, CCIS proceedings are included in the CCIS electronic book series hosted in the SpringerLink digital library at http://link.springer.com/bookseries/7899. Conferences publishing in CCIS are allowed to use our online conference service (Meteor) for managing the whole proceedings lifecycle (from submission and reviewing to preparing for publication) free of charge.

Publication process

The language of publication is exclusively English. Authors publishing in CCIS have to sign the Springer CCIS copyright transfer form, however, they are free to use their material published in CCIS for substantially changed, more elaborate subsequent publications elsewhere. For the preparation of the camera-ready papers/files, authors have to strictly adhere to the Springer CCIS Authors' Instructions and are strongly encouraged to use the CCIS LaTeX style files or templates.

Abstracting/Indexing

CCIS is abstracted/indexed in DBLP, Google Scholar, EI-Compendex, Mathematical Reviews, SCImago, Scopus. CCIS volumes are also submitted for the inclusion in ISI Proceedings.

How to start

To start the evaluation of your proposal for inclusion in the CCIS series, please send an e-mail to ccis@springer.com

Aswin Kannan · Sachit Rao · Chitra Annamalai
Editors

Advanced Computing and Communications: Technologies for a Sustainable Future

30th Advanced Computing and Communications Conference, ADCOM 2025
Bangalore, India, December 17–19, 2025
Proceedings

Editors
Aswin Kannan
International Institute of Information Technology Bangalore
Bengaluru, Karnataka, India

Sachit Rao
International Institute of Information Technology Bangalore
Bengaluru, Karnataka, India

Chitra Annamalai
PSG College of Technology
Coimbatore, Tamil Nadu, India

ISSN 1865-0929 ISSN 1865-0937 (electronic)
Communications in Computer and Information Science
ISBN 978-3-032-26268-4 ISBN 978-3-032-26269-1 (eBook)
https://doi.org/10.1007/978-3-032-26269-1

© The Editor(s) (if applicable) and The Author(s), under exclusive license to Springer Nature Switzerland AG 2026

This work is subject to copyright. All rights are solely and exclusively licensed by the Publisher, whether the whole or part of the material is concerned, specifically the rights of translation, reprinting, reuse of illustrations, recitation, broadcasting, reproduction on microfilms or in any other physical way, and transmission or information storage and retrieval, electronic adaptation, computer software, or by similar or dissimilar methodology now known or hereafter developed.
The use of general descriptive names, registered names, trademarks, service marks, etc. in this publication does not imply, even in the absence of a specific statement, that such names are exempt from the relevant protective laws and regulations and therefore free for general use.
The publisher, the authors and the editors are safe to assume that the advice and information in this book are believed to be true and accurate at the date of publication. Neither the publisher nor the authors or the editors give a warranty, expressed or implied, with respect to the material contained herein or for any errors or omissions that may have been made. The publisher remains neutral with regard to jurisdictional claims in published maps and institutional affiliations.

This Springer imprint is published by the registered company Springer Nature Switzerland AG
The registered company address is: Gewerbestrasse 11, 6330 Cham, Switzerland

If disposing of this product, please recycle the paper.

Preface

The 30th International Conference on Advanced Computing and Communications (ADCOM 2025), the flagship systems conference of the Advanced Computing and Communications Society (ACCS), was held at the International Institute of Information Technology Bangalore (IIIT Bangalore), India, from December 17 to 19, 2025. ADCOM is a premier annual international conference that brings together researchers, practitioners, and students from academia and industry to discuss advances in computational and communication systems.

The theme of ADCOM 2025, "Technologies for a Sustainable Future," reflects the growing importance of developing environmentally responsible and sustainable computing solutions. The conference provided a platform for presenting and discussing cutting-edge research addressing challenges in areas such as optimization, machine learning, cloud computing, robotics, intelligent systems, and sustainable technologies.

A total of 76 papers were submitted to the conference, including 72 regular (long) papers and 4 PhD forum (short) papers. The review process was conducted in a rigorous double-blind manner using the EasyChair system, with a total of 28 reviewers participating in the evaluation process. Of the submissions, 60 papers underwent full review, and each received two independent reviews. The remaining 16 submissions were pre-screened by at least two members of the Program Committee and were not considered for full review based on quality considerations. Based on the reviews and subsequent discussions, 13 regular papers and 3 PhD forum papers were accepted for presentation and inclusion in these proceedings. There were no invited papers.

To provide a coherent structure, the accepted papers have been organized into thematic sections. The section "Computation, Control, and Automation" includes contributions on energy-aware computing, cloud resource optimization, intelligent surveillance, robotic exploration, and signal-based flight analysis. The section "Energy and Environment" presents work on renewable energy systems, solar potential estimation, wireless energy efficiency, and urban sustainability modeling. The section "Machine Learning and Natural Language Processing" highlights advances in semantic intelligence, generative AI, multilingual systems, and healthcare-related machine learning, including both applied frameworks and survey contributions. In addition, the proceedings include a "PhD Forum" section featuring early-stage research on explainable AI for surveillance, environmental health risk prediction, and efficient visual question answering systems.

We would like to express our sincere gratitude to the members of the Program Committee and the reviewers for their valuable time and insightful feedback, which ensured the high quality of the selected papers. We also thank all the authors for their contributions and for making ADCOM 2025 a success.

Aswin Kannan
Sachit Rao
Chitra Annamalai

Organization

General Chairs

Srinath Srinivasa	IIIT Bangalore, India
Saragur Srinidhi	Prometheus Consulting, ACCS, India

Organizing Chair

Jailendra Kumar	HighNorm Systems and Services Pvt. Ltd., ACCS, India

Program Committee Chairs

Aswin Kannan	IIIT Bangalore, India
Sachit Rao	IIIT Bangalore, India
Chitra Annamalai	PSG College of Technology, India

Program Committee

Sushree Behera	IIIT Bangalore, India
Vaishnavi Gujjula	IIIT Bangalore, India
Ahana Pradhan	IIIT Bangalore, India
Tulika Saha	IIIT Bangalore, India
Ramasuri Narayanam	Adobe Research, India
Amar Prakash Azad	Fujitsu Research, India
Palanivel Kodeswaran	Walmart, India

Additional Reviewers

(*IIIT Bangalore, India; WIAS-Berlin, Germany; and HU-Berlin, Germany*)

Naganand Yadati	Aswathi Mundayatt Valappil
Karthikeya Subramanian	Sachin Mishra
V. Ramasubramanian	Patenge Krishna

Monish K.
Meenakshi D'Souza
Pooja Bassin
Shri Tadinada
Amal Alphonse
Rahisha Thottolil
Divyam Sareen
Navitha Parthasarathy
Irfan Mohammad
Kausthubh Manda
Me Me Khaing
Viswanath Gopalakrishnan
Sayak Choudhury
Mahima B.
Zijun Li

Contents

Computation, Control, and Automation

Quantitative Evaluation of Software Carbon Intensity for Serverless Applications 3
Mahima and Meenakshi D'Souza

A Hybrid Genetic Algorithm with Restricted Mutation (GA-RM) for Task Scheduling in Cloud Computing 23
Gudimetla Lakshmidhara Chandan, Abhinav Kumar Singh, K. Bala Trishank, and Vaidehi Vijayakumar

Multi-modal Automatic Detection of Distress Behaviors in Surveillance Systems 36
K. Gayathri, A. Chitra, and S. Priya Dharshini

EAMO: An Algorithm for Exploration Among Movable Objects Using a Ground Robot 51
Nitin Kumar Dhiman, Aniket Kalra, and Deepak Khemani

Hybrid Method for Automatic Flight Manoeuvre Recognition Using Signal Processing and Classification Learner 66
Akash Sahu, Khadeeja Nusrath, and Dushyant Kaliyari

Energy and Environment

Net Zero Energy Home with Solar Powered Photovoltaic System 87
Makarand M. Lokhande, Prakash S. Kulkarni, Pravin D. Sawarkar, Ramsha Karampuri, Anindita Roy, and Raju Lenkalapelly

Automated Assessment of Rooftop Solar Energy Potential Using Deep Learning-Driven Building Footprint Extraction Model 99
Sachin Mishra

Analysis of Energy Efficiency of a Distributed STAR-RIS-Assisted Wireless System 115
H. Rashmi, Ashvini Chaturvedi, and Bodempudi Naga Siva Prasad

Multi-task Graph Convolutional Network Framework for Ward-Level Urban Sustainability in Bengaluru 130
C. A. Prajwal and K. P. Impana

Machine Learning and Natural Language Processing

TamilBookAnnot: A Strategic e-Book Annotation Framework in Tamil Integrating Semantic Intelligence Interfaced with Knowledge 147
S. A. Mohammed Salman, Gerard Deepak, and A. Santhanavijayan

AssameseDocRec: A Strategic Framework for Assamese Document Recommendation Integrating Generative Semantic Intelligence and Quantitative Semantic Reasoning 160
Anubrat Bora, Gerard Deepak, and Samiksha Shukla

AssameseVidRec: A Strategic Framework for Assamese Video Recommendation Integrating Generative Semantic Intelligence 173
Anubrat Bora and Gerard Deepak

A Comprehensive Survey of Traditional, Machine Learning, Deep Learning, and Federated Learning Methods for Multimodal Autism Spectrum Disorder Diagnosis 183
A. Sathish Kumar, Malmathanraj Ramnathan, Avik Hati, and P. Palanisamy

PhD Forum

An Explainable Deep Learning Framework for Suspicious Activity Recognition and Person Identification in Video Surveillance 201
K. Gayathri and A. Chitra

Prediction of Respiratory and Cardiovascular Health Risks Due to Air Pollutants Using Similarity Measures 215
Neha Bhatt and Asha Joseph

PhD Forum 2025 - Development of Interpretable and Efficient Visual Question Answering Models 228
Souvik Chowdhury and Badal Soni

Author Index 239

Computation, Control, and Automation

Quantitative Evaluation of Software Carbon Intensity for Serverless Applications

Mahima(✉) and Meenakshi D'Souza

International Institute of Information Technology Bangalore, Bengaluru, Karnataka, India
{Mahima.B,meenakshi}@iiitb.ac.in

Abstract. Serverless computing is becoming popular in cloud software development because of its scalability, less management effort and pay as-you-go features. But the high abstraction hides the hardware details making it difficult to measure the environmental impact, particularly the Software Carbon Intensity (SCI) of such applications. In this paper, we propose a simple framework to measure SCI for serverless workflows hosted on AWS Cloud, using the SCI Specification guidelines. Our methodology calculates the energy consumption for each resource by pulling metrics from AWS CloudWatch and Lambda Insights. Using a case study of a sample workflow, we show the calculations involved in computing the energy consumption of different resources and also the total carbon emissions. Our results suggest that since 48% of the parameters are unknown, serverless may not be the best choice for developers aiming to build greener software.

Keywords: Serverless · Software Carbon Intensity · Sustainability · Cloud Computing

1 Introduction

Sustainability in computing has emerged as a pressing concern due to rapid increase in ICT devices. With the expansion of the cloud network, the demand for flexible and scalable platforms has increased significantly. Among these, serverless computing has gained momentum because of its pay-per-use model, auto-scaling, and operational simplicity. However, while serverless offers economic and development benefits, the environmental impacts remain underexplored. Understanding the energy consumption and associated carbon emissions in serverless applications is therefore crucial to ensure shifting towards cloud-native architecture aligns with the goals of sustainability.

Recently, the International Standard Organization (ISO) released the standard ISO/IEC 21031:2024 Software Carbon Intensity Specification [4] which describes the methodology to calculate the SCI score of a software system. This

© The Author(s), under exclusive license to Springer Nature Switzerland AG 2026
A. Kannan et al. (Eds.): ADCOM 2025, CCIS 2947, pp. 3–22, 2026.
https://doi.org/10.1007/978-3-032-26269-1_1

recognition emphasizes that sustainability is not only a secondary attribute, but a measurable aspect of modern software engineering.

An important aspect of the SCI specification [2] is the choice of functional unit R which describes how the application scales. The specification provides the flexibility in selecting R depending on the context of the system under study. Suggested list of functional units includes an API call or request, a benchmark, an user, a machine, a minute/time unit, a device, a physical site, data volume, a batch or scheduled job, a transaction, or a database read/write, providing an exhaustive list of possible granularity in which software carbon intensity can be measured.

The main goal of this paper is to develop metrics and methodologies to calculate the carbon intensity of software applications running on a serverless architecture. The abstraction inherent in serverless platforms hides the hardware details, making it difficult for both practitioners and researchers to get access to parameters needed for precise calculation of the energy usage and carbon intensity of serverless workflows. The absence of standardized methodologies leaves a significant gap in both theory and practice. While it is easy to calculate the SCI for one executing method or function, extending it to workflows spanning across a serverless architecture is non-trivial [1].

The following is a list of results presented in this paper:

- A **quantitative evaluation of feasible SCI measurement** of serverless workflows by using monitoring tools available within serverless workflows.
- Identifying and categorizing the **known and unknown parameters** that influence the sustainability measurement.
- Discussing the **trade-offs between abstraction and sustainability**, offering the limitations of working with serverless environments as far as sustainability is concerned.
- Demonstrating how dashboards and visualization techniques can be used to **aggregate, display, and interpret sustainability metrics**.
- Providing a **prototype for calculating SCI** for serverless applications within the AWS platform.

The paper is organized as follows. The next section introduces the notion of sustainability in computing, defines the metric of Software Carbon Intensity (SCI). We discuss related work focusing on sustainability in software and computing applications in Sect. 3. We focus on serverless computing which is described in Sect. 4, followed by our benchmark case study example developed on the AWS platform in Sect. 5. The SCI calculations for workflows of serverless applications are done in Sect. 6 and the experiments on the case study are illustrated in Sect. 8. Finally, we conclude and describe the on-going work in the last two sections.

2 Background

2.1 Sustainability in Computing

Sustainability in computing is not just about reducing the costs or optimizing resource allocation. It has broader environmental implications since data centers emit large amounts of carbon. The Greenhouse Gas Protocol provides guidance standards to practitioners who carry out assessments of ICT products. The GREENSOFT reference model [1] divides the carbon emissions coming from software into three scopes:

- Scope 1 (Direct emissions): Emissions produced directly from sources that the company owns or controls, such as fuel combustion in company vehicles, boilers, and process equipment.
- Scope 2 (Indirect emissions from purchased energy): Emissions from the generation of electricity, steam, heating, or cooling that the company purchases and consumes, occurring at the energy producer's facility.
- Scope 3 (Other indirect emissions): Emissions that result from the company's activities but happen from sources it does not own or control.

2.2 Software Carbon Intensity

In this study, we apply the methodology specified by the SCI specification document [2]. Towards calculating and reporting the SCI score, the methodology begins with defining the software boundary, i.e., identifying the system components to be included in the analysis. Next, the functional unit that best presents how the application scale is selected as the SCI is expressed as the rate of carbon emission per functional unit. For each component within this boundary, quantification method is chosen - either real world based measurements or lab-based model estimations. The next step is to quantify the rate for each component and take an aggregate of SCI values of all these components in the system. Finally, the SCI score is reported along with the defined boundary and the calculation methodology.

SCI specification document classifies the emissions into Operational and Embodied Emissions. Operational Emissions is the product of energy consumed by the hardware the software is running on and the region-specific carbon intensity. The energy consumption in this case includes all energy consumed by hardware or provisioned. The region-specific carbon intensity [17] measures the grid carbon intensity of the grid region. Embodied emissions refer to the carbon emissions generated during the manufacturing, transportation, and eventual disposal of the hardware on which software runs. However, in the case of serverless computing, the underlying hardware is fully managed and abstracted by the cloud provider. Since users neither own nor directly interact with the hardware, and because reliable data on the life cycle emissions of this hidden infrastructure is not accessible, this study focuses only on operational emissions. Therefore, embodied emissions are excluded from our analysis.

3 Related Work

Research on measuring Software Carbon Intensity has gained increasing attention in recent years. The Software Carbon Intensity (SCI) specification [2] provides a standardized framework to express the carbon emissions by normalizing it with the functional unit R. SCI has been used in both industrial and academic contexts as a guideline for reporting software-related emissions. Complementary to SCI, the GHG Protocol [3] also focuses on estimating the energy consumption particularly in cloud hosted applications. Several studies have examined sustainability in cloud computing. Cloud Providers, like AWS, have been successful in developing their custom tool to calculate the carbon footprint of the applications [16] hosted on the cloud. The Cloud Carbon Footprint (CCF) [5] methodology provides an open source, transparent approach to measure, monitor and reduce carbon emissions from the cloud. The framework focuses on estimating the emissions associated with standalone functions within a cloud framework. A similar work was carried out by Awwad et al. [6] who estimated the carbon footprint of serverless functions on a public cloud platform. Their study provides valuable insights into the energy and emission characteristics of serverless workloads but they are restricting it to analyzing standalone functions. Our work complements this line of research by applying the SCI framework to the entire workflow and demonstrating the uncertainties and tradeoffs that arise when examining the sustainability in the context of serverless frameworks. While prior work primarily focuses on standalone Lambda functions, we provide a metric-driven methodology for workflow-level SCI estimation in serverless workflows and characterize how missing parameters limit the accuracy of the resulting carbon emission calculations.

4 Serverless Computing Overview

4.1 Serverless Architecture

Serverless Computing or Function-as-a-service (FaaS) is a cloud computing model where developers write and deploy small, independent code functions that execute in response to event triggers without having to manage the underlying server infrastructure. The cloud provider automatically handles hardware, operating systems, and web servers allowing the developers to focus solely on the code. FaaS approach charges users based on actual function execution time and consumption of (hardware) resources. In this paper, we consider Amazon Web Services (AWS) [16] as a cloud provider service for all our experiments. One of the major use cases of this model is IoT data processing. An IoT framework needs a computing model where the appropriate events are triggered based on the responses received from the connected to other devices or storage. We explain the key components of AWS below. We also would like to note that the SCI calculation framework proposed in this paper applies for other serverless frameworks too, AWS is used as a case study only.

4.2 Key Components

AWS offers serverless resources tailored for various use cases such as code execution, object storage, databases, etc., that eliminate the need for explicit infrastructure configuration by developers. In this experiment, we have considered the following AWS Resources.

- **AWS Lambda**: An event-driven compute service that allows users to run their code without provisioning or managing servers [9].
- **AWS Simple Storage Service (S3)**: Scalable Object Storage Service offered by Amazon [10].
- **Amazon DynamoDB**: Serverless NoSQL Database service offered by AWS [11].
- **Amazon API Gateway**: Fully managed service provided by AWS that acts as a "front door" for applications to access and transmit data, functionality, etc., from various backend services like AWS Lambda [12].
- **Amazon Rekognition**: Cloud-based AI service provided by AWS that enables the addition of image analysis capabilities to applications. It leverages deep learning technologies that do not require any machine learning expertise from the user [13].
- **Amazon CloudWatch**: Used to collect and track metrics and logs. It can also be used to visualize the performance of the applications by creating custom dashboards in it [14].

4.3 Advantages and Limitations

Serverless architecture offers advantages like automatic scalability, faster development cycle by eliminating the need for server management and reduced operational costs with their pay-as-you-go model. However, it has its limitations including limited control over infrastructure, challenges with cold start, and debugging and testing challenges.

5 Case Study: Serverless Home Security Application

5.1 Application Overview

The application presented in this work is a serverless home security system designed and deployed on Amazon Web Services (AWS). It integrates multiple cloud-native services to achieve different use cases like storage, database and AI technology. By combining API Gateway, Lambda, DynamoDB, S3, and Rekognition, the application eliminates the need for traditional infrastructure while ensuring scalability and cost-effectiveness. This application is particularly relevant in the context of contemporary smart homes and apartment complexes, where controlled access and user safety are critical concerns.

5.2 Application Workflow

The proposed home security application integrates multiple serverless components to enable secure and efficient access control. The workflow begins when a user interacts with the doorbell system, which triggers a logging event through the DoorbellLambda function. This event records the house identifier (e.g., H1) into the DoorbellEvents DynamoDB table, ensuring that all entry attempts are tracked in real time.

Once the doorbell is activated, the next phase involves image upload and face recognition. The user uploads an image, which is stored in the designated S3 bucket under the uploads folder. This upload event automatically invokes the FaceRecognitionLambda, which retrieves the latest doorbell event from DynamoDB and compares the uploaded image against the pre-registered dataset stored in the corresponding household folder within S3. For this process, the application leverages the Amazon Rekognition service to perform facial matching. If a valid match is detected, the associated User ID is retrieved from the HomeSecurityUsers DynamoDB table.

The final stage of the workflow involves PIN-based verification to strengthen security. A verification interface prompts the user to enter a PIN, which, along with the identified User ID, is sent to the PINVerificationLambda. This function cross-checks the entered PIN against the plain-text PIN stored in DynamoDB. If the match is successful, the system grants access; otherwise, the request is denied.

The application components thus consist of:

- Five Lambda functions responsible for website fetching, doorbell logging, upload handling, facial recognition, and PIN verification.
- Five API Gateway endpoints that expose secure interaction points for system functions.
- Two S3 buckets, one dedicated to user image storage and another for hosting static website content.
- Two DynamoDB tables for recording doorbell events and maintaining user credentials.
- Amazon Rekognition, which provides facial recognition as a managed AI service.

5.3 Workflow Illustration

The workflow represents a sequence of connected cloud-based services executing the serverless application logic. Figure 1 illustrates the components and data flow involved. Each step in the workflow corresponds to function invocations, storage operations, or API interactions that together realize the application objective.

In the next section, we detail the SCI-based methodology used to quantify the energy consumption and carbon emissions for this workflow.

6 Methodology

6.1 SCI Formulae and Definitions

In the Green Software Foundation (GSF) documentation, Software Carbon Intensity (SCI) [2] is defined as the rate of carbon emissions per functional

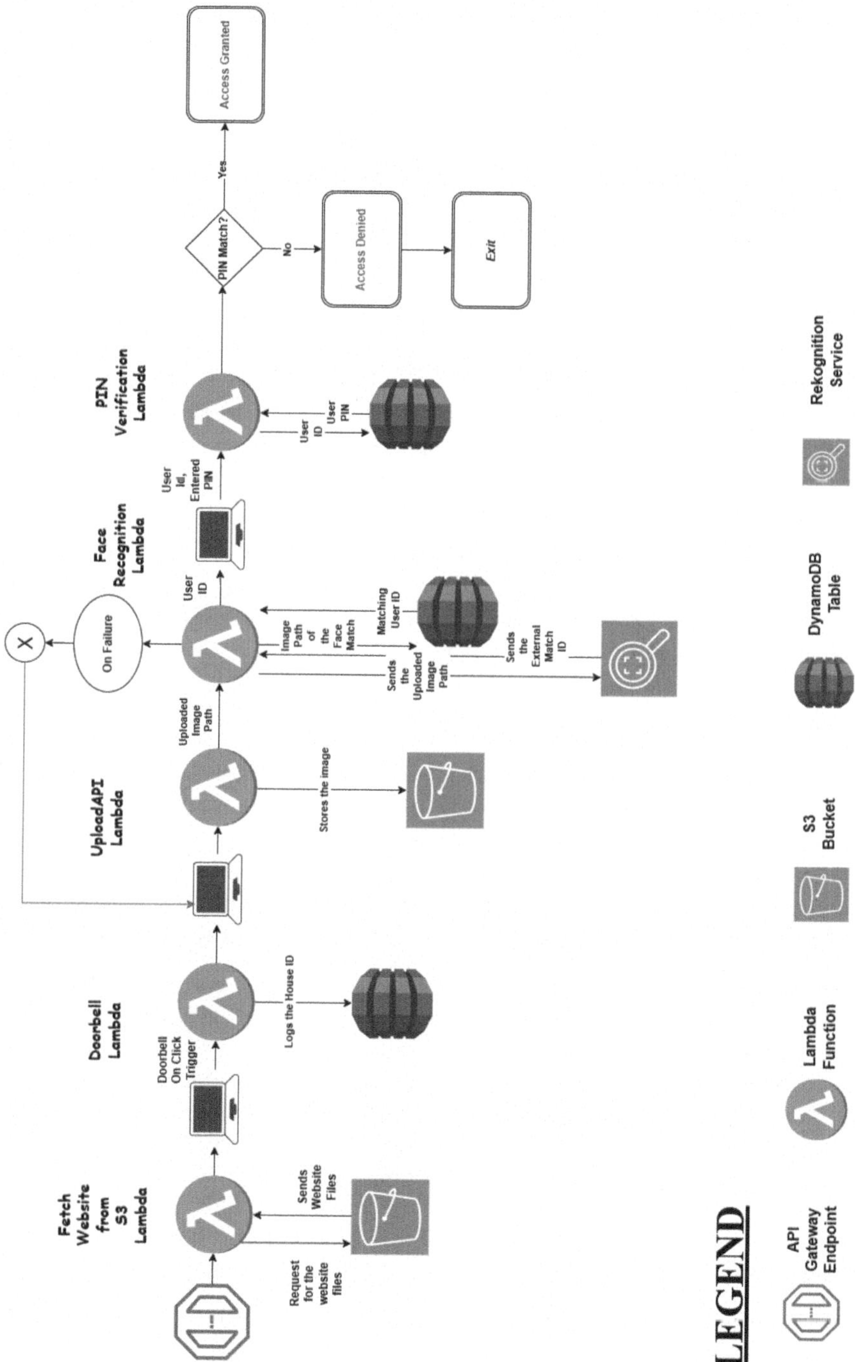

Fig. 1. Workflow diagram illustrating the serverless application components and data flow.

unit, denoted as R. For this study, we consider the functional unit to be one invocation of the application workflow—that is, one complete execution of the workflow.

SCI represents the sum of both operational emissions and embodied emissions. However, without loss of generality, and given the nature of serverless applications where hardware is abstracted, we safely omit embodied emissions from our analysis.

Operational Emissions associated with a software application is given as a product of energy consumption of the hardware the software is running on and the region-specific carbon intensity.

$$O = E \times I \tag{1}$$

where E is the energy consumed by a functional unit of work (e.g., execution of code, database, etc.) in kWh, and I is the region-specific carbon intensity of the region in which the application runs, measured in gCO_2/kWh.

The basic physics formula for energy is given by:

$$Energy = Power \times Time \tag{2}$$

We apply this formula to calculate the energy consumption of each resource utilized by the application.

6.2 Workflow Layered Approach

In order to estimate the Software Carbon Intensity of the serverless workflow, we adopt a layered approach. The overall SCI is decomposed stepwise, starting from the atomic level components such as compute, memory, storage, and network resources and then aggregating these component-level emissions to derive the composite workflow emission. For each functional unit of the workflow, we calculate the energy consumption across the following categories:

- **Compute Energy Consumption:** Emissions resulting from the energy consumed by the central processing unit (CPU) while executing the software functions.
- **Memory Energy Consumption:** Emissions attributable to the energy used by memory components, including DRAM and ephemeral memory, during both initialization and execution phases of the software.
- **Storage Energy Consumption:** Emissions arising from the energy consumed by storage devices responsible for holding application data.
- **Network Energy Consumption:** Emissions generated through the energy expenditure required in transmitting data over the network infrastructure.

6.3 Data Source and Collection Methods

We have collected metrics and logs from CloudWatch and Lambda Insights [15] as sources. These metrics are segregated into categories based on the type of emissions they contribute to. Table 1 lists the collected metrics along with their source, associated resource, and units. In general, when a serverless application executes within the serverless platform, all of these information can be collected as data and used in our proposed framework.

Table 1. Metrics collected from CloudWatch and Lambda Insights

Metric Name	Unit	Source
CPU Total Execution Time	milliseconds	Lambda Insights
Memory Allocated	MB	Lambda Insights
Maximum Memory Used	MB	Lambda Insights
Temporary Memory Used	Bytes	Lambda Insights
Execution duration	milliseconds	Lambda Insights
Init Duration	milliseconds	Lambda Insights
Total Data Transferred	Bytes	Lambda Insights
AllRequests	Count	CloudWatch S3 metrics
TotalRequestLatency	milliseconds	CloudWatch S3 metrics
BucketSizeBytes	Bytes	CloudWatch S3 metrics
SuccessfulRequestLatency	milliseconds	CloudWatch DynamoDB metrics
Latency	milliseconds	CloudWatch API Gateway metrics
Response Time	milliseconds	CloudWatch Rekognition Service metrics

6.4 Energy and Emission Calculations Using Measurable Metrics

In this work, we adopt a bottom-up approach to estimate the energy usage and carbon emissions of our serverless workflow. The methodology begins by calculating the energy consumed by individual components such as function execution, data transfer, and storage. These component-level values are then aggregated to obtain the total energy consumption of the workflow, which is subsequently translated into carbon emissions using region-specific carbon intensity factors. The following equations summarize this process.

1. Total Energy Consumption

$$E_{\text{Total}} = E_{\text{Lambda}} + E_{\text{Storage}} + E_{\text{DB}} + E_{\text{API Gateway}} + E_{\text{Rekog}} \tag{3}$$

where

- E_{Total} is the total operational energy consumption (in kWh)
- E_{λ} is the energy consumed by a computation function (in our prototype, AWS Lambda)
- E_{Storage} is the total energy consumption by a storage component (in our prototype, Amazon S3)

- E_{DB} is the total energy consumption by a relational database resource (in our prototype, DynamoDB)
- $E_{\text{API Gateway}}$ is the energy consumed by API Gateway
- E_{Rekog} is the energy consumed by the Visual Recognition Source (in this prototype, Amazon Rekognition service - a Deep Learning Model for visual recognition).

2. Lambda Energy Consumption: For the Lambda resource, the total energy consumption is calculated by aggregating the contributions from all relevant categories, namely compute, network, and storage. This bottom-up breakdown is expressed as:

$$E_{Lambda} = E_{L_{compute}} + E_{L_{memory}} + E_{L_{storage}} + E_{L_{network}} \quad (4)$$

where the individual components are defined as follows:

- **Lambda × Compute Energy Consumption:** Each Lambda function is allocated compute proportional to its configured memory. A Lambda function with ˜1769 MB of memory is allocated approximately one vCPU. We model the energy consumed by the CPU during the active execution time of the Lambda function.

$$E_{L_{compute}} = \frac{T_{\mathrm{cpu}} \times P_{\mathrm{core}} \times M_{\mathrm{alloc}}}{1000 \times 3600 \times 1000 \times M_{\mathrm{ref}}} \quad (5)$$

  - $E_{L_{compute}}$ = Energy consumed for compute (in kWh)
  - T_{cpu} = Total CPU execution time (in milliseconds)
  - P_{core} = Power rating of a single CPU core (in watts)
  - M_{alloc} = Memory allocated to the Lambda function (in MB)
  - M_{ref} = 1769 MB (reference memory that maps to 1 vCPU) **Note:** Constants used for unit conversion:

  - 1000 × 3600 × 1000 converts milliseconds × watts to kilowatt-hours.

  The color coding is used to distinguish the parameters that can be measured (green) from the ones that cannot be measured (red).

- **Lambda × Memory Energy Consumption:** Lambda uses certain amount of memory coming from the memory element to perform one operation throughout the init phase and actual execution.

$$E_{L_{memory}} = \frac{M_{\mathrm{used}} \times P_{\mathrm{mem}} \times (T_{\mathrm{exec}} + T_{\mathrm{init}})}{1024 \times 1000 \times 1000 \times 3600} \quad (6)$$

  - $E_{L_{memory}}$ = Energy consumed for memory (in kWh)
  - M_{used} = Maximum memory used (in MB)
  - P_{mem} = Power rating of DRAM memory (in watts per GB)
  - T_{exec} = Execution duration (in ms)
  - T_{init} = Initialization duration (in ms)
  Note: Unit conversions used:

- 1024: Convert MB to GB
- $1000 \times 3600 \times 1000$ converts milliseconds $\times$ watts to kilowatt-hours

– **Lambda × Storage Emissions:**
  - AWS Lambda provides ephemeral storage space which is local to the Lambda environment and provisioned through the entire lifecycle.

$$E_{L_{storage}} = \frac{S_{\text{used}} \times P_{\text{storage}} \times (T_{\text{exec}} + T_{\text{init}})}{1024^3 \times 1000 \times 1000 \times 3600} \tag{7}$$

  - $E_{L_{storage}}$ = Energy consumed by ephemeral storage (in kWh)
  - S_{used} = Ephemeral storage used (in Bytes)
  - P_{storage} = Power rating of NVMe SSD (in watts per GB)
  - T_{exec} = Execution duration (in ms)
  - T_{init} = Initialization duration (in ms)
    Note: Unit conversions used:
  - 1024^3: Convert Bytes to GB
  - $1000 \times 3600 \times 1000$ converts milliseconds $\times$ watts to kilowatt-hours

– **Lambda × Network Energy Consumption:** Network emissions are calculated by multiplying the energy intensity factor (kWh/GB) by the amount of data transferred; only the volume of data matters, not the time taken.

$$E_{\text{network}} = \frac{D_{\text{network}} \times \epsilon_{\text{network}}}{1024^3} \tag{8}$$

  - E_{network} = Energy consumed during network transfer (in kWh)
  - D_{network} = Total data transferred (in Bytes)
  - $\epsilon_{\text{network}}$ = Energy intensity of the network (in kWh/GB)
    Note: Unit conversions used:
  - 1024^3: Convert Bytes to GB

3. Storage component's Energy Consumption: Amazon S3 is Amazon's storage service for storing objects. Emissions from computation, memory, and network are handled by Lambda, so S3 emission calculations focus solely on storage. These are categorized into **active power consumption** (from SSD activity during requests) and **idle power consumption (from storing objects)**. Network-related emissions occur during data transfer but are accounted for within Lambda. The total energy consumption is expressed as follows:

$$E_{Storage} = E_{S3_{Active}} + E_{S3_{Idle}} \tag{9}$$

where the individual components are defined as follows:

– **S3 Active Energy Consumption:** Active energy consumption accounts for the energy used when performing storage operations such as object uploads, downloads, and retrievals.

$$E_{S3_{active}} = \frac{R_{\text{total}} \times D_{\text{latency}} \times P_{\text{active}}}{1000 \times 3600 \times 1000} \tag{10}$$

- $E_{S3_{active}}$ is the energy consumed during active operation of S3 (in kWh)
- R_{total} is the total number of S3 requests
- D_{latency} is the total request latency (in milliseconds)
- P_{active} is the power drawn by active storage element (in watts)
  Note: Constants used for unit conversion:
- $1000 \times 3600 \times 1000$ converts milliseconds × watts to kilowatt-hours

– **S3 Idle Energy Consumption:** Idle energy consumption reflects the baseline energy required to maintain stored data even when no active read/write operations occur. This is modeled as a function of the data volume stored and the idle energy intensity factor:

$$E_{S3_{idle}} = \left(\frac{B + (N_{\text{inv}} \times S_{\text{upload}})}{1024^3}\right) \times \left(\frac{P_{\text{idle}}}{C_{\text{mem}} \times 1000}\right) \times 1 \tag{11}$$

- $E_{S3_{idle}}$ is the idle energy consumption of Amazon S3 (in kWh)
- B is the current size of the S3 bucket (in bytes)
- N_{inv} is the number of invocations (uploads)
- S_{upload} is the size of a single upload (in bytes)
- P_{idle} is the idle power of the storage element (in watts)
- C_{mem} is the capacity of the storage element (in GB)
- 1024^3 converts bytes to GB
- 1 represents the 1-h time duration (in hours)
- 1000 converts watts to kilo-watts

Amortized over 1 h.

4. DB Energy Consumption: Energy consumption is based on the database operations per invocation and follows the same concept as S3. Estimating idle power requires the table size and usage metrics, but since CloudWatch does not provide table size, idle power calculations are deferred. The calculations are given as:

$$E_{\text{DB}} = \frac{L_{\text{success}} \times P_{\text{active}}}{1000 \times 3600 \times 1000} \tag{12}$$

- E_{DB} is the active energy consumption of DynamoDB (in kWh)
- L_{success} is the total latency of successful requests (in milliseconds)
- P_{active} is the active power consumption of the storage element (in watts)
- $1000 \times 3600 \times 1000$ converts milliseconds × watts to kilowatt-hours

5. API Gateway Energy Consumption: Emissions originate from the underlying server hardware, mainly compute. Network emissions are accounted for by Lambda, while compute emissions depend on the processor's power rating. Storage and Memory emissions are not applicable.

$$E_{\text{API Gateway}} = \frac{L_{\text{API}} \times P_{\text{Proc}}}{1000 \times 3600 \times 1000} \tag{13}$$

- $E_{\text{API Gateway}}$ is the energy consumed by the API Gateway (in kWh)

- L_{API} is the total latency of API Gateway requests (in milliseconds)
- P_{Proc} is the Power Rating of the underlying processor (in watts)
- $1000 \times 3600 \times 1000$ converts milliseconds $\times$ watts to kilowatt-hours

6. Rekognition Energy Consumption: Emissions are solely from compute, as all deep learning models run on GPUs whose power consumption (TDP) determines the emissions.

$$E_{\text{Rekog}} = \frac{T_{\text{response}} \times P_{\text{GPU}}}{1000 \times 3600 \times 1000} \tag{14}$$

- E_{Rekog} is the energy consumed by the Rekognition service (in kWh)
- T_{response} is the total response time of the Rekognition API (in milliseconds)
- P_{GPU} is the power consumption of the GPU used (in milliwatts)
- $1000 \times 3600 \times 1000$ converts milliseconds $\times$ watts to kilowatt-hours

7. Final Carbon Emissions: The Carbon Emissions for the entire workflow of the application can be computed as:

$$SCI = E_{\text{Total}} \times PUE \times GridEmissionFactor \tag{15}$$

- SCI is the Software Carbon Intensity of the Application per invocation (in gCO_2)
- E_{total} is the total operational energy consumption (in kWh)
- PUE is the Power Usage Effectiveness of the data centre
- GridEmissionFactor is the Regional Carbon Intensity of the region where the application runs (in gCO_2/kWh)

7 Trade-Offs in Measurement

7.1 Known v/s Unknown Parameters

Since serverless architectures abstract away the underlying infrastructure, cloud providers such as AWS do not disclose hardware-level details like power ratings. As a result, nearly 48% of the parameters required for accurate sustainability measurement remain unavailable and must instead be assumed. This limitation directly affects the accuracy of Software Carbon Intensity (SCI) calculations, as some of the variables in the energy and emission formulas cannot be derived from CloudWatch metrics. Table 2 summarizes the parameters that can be directly obtained from monitoring tools versus those that remain unmeasurable.

7.2 Abstraction v/s Sustainability

The tradeoff between Abstraction and Sustainability creates a fundamental tension. One of the defining advantages of Serverless architectures is the high level of abstraction it provides. Amazon itself acknowledges the current absence of

Table 2. Measured vs. Assumed/Unavailable Parameters

Measured Parameters	Assumed/Unavailable Parameters
CPU total execution time (T_{cpu})	Power rating of CPU core (P_{core})
Memory allocated ($Malloc$)	Power rating of DRAM memory (P_{mem})
Maximum memory used (M_{used})	Power rating of NVMe SSD ($P_{storage}$)
Temporary storage used (S_{used})	Energy intensity of network ($\varepsilon_{network}$)
Execution duration (T_{exec})	Power consumption of GPU (P_{GPU})
Initialization duration (T_{init})	Power consumption of processor used in API Hardware (P_{Proc})
Total data transferred ($D_{network}$)	Active/Idle power of S3 memory elements (P_{active}, P_{idle})
Total S3 requests (R_{total})	Capacity of memory element (C_{mem})
S3 total request latency ($D_{latency}$)	Grid emission factor
S3 bucket size (B)	Power Usage Effectiveness (PUE)
DynamoDB latency ($L_{success}$)	Number of uploads (N_{inv})
API Gateway latency (L_{API})	Size of single upload (S_{upload})
Rekognition response time ($T_{response}$)	

a dedicated sustainability lens for serverless offerings, highlighting the inherent difficulty in precisely measuring and managing carbon emissions in such environments. [7] While this abstraction accelerates the development work, it creates a significant trade-off with sustainability. The same parameters such as power ratings of processors, memory or storage devices necessary for the computation of Carbon Emissions of a software application are concealed by the serverless service providers. As a result, sustainability practitioners must rely on assumed constants which can lead to uncertainty in the final emission value. This tension between abstraction and sustainability highlights the need for better architectures as far as green software is concerned.

8 Results

8.1 Quantitative Impact Analysis

To better understand the impact of unavailable parameters, we substituted assumed constants in place of missing hardware-level values such as power ratings and grid emission factors. The results indicate that nearly half of the variables required for accurate measurement must be approximated. When these variations are added up across the whole serverless workflow, the errors accumulate, making it difficult to calculate SCI accurately.

8.2 Sample Calculations

To illustrate the methodology, we present the detailed calculation for the Lambda function that provides face recognition, which involves CPU, memory, network, and storage activities. The total energy consumption E_{total} is the sum of the energy consumed by each resource.

Table 3 presents the values assigned to the assumed parameters in the formula, along with the corresponding sources from which these values are derived.

Notes

- P_{core}: Experiments utilized AWS Graviton2 processors (64 cores), with total power consumption assumed at 80–110 W based on available benchmarks. An average of 95 W yields $P_{\text{core}} = 1.48$ W/core for face recognition workloads [18].
- P_{storage}: Represents the power rating of ephemeral storage specific to Lambda functions.
- $P_{\text{S3, active}}$, $P_{\text{S3, idle}}$, P_{DDB}: SESAME '25 identifies SSD usage in AWS data centers, while AWS documentation confirms NVMe for EBS volumes. Average active power of 18.5 W applied per SNIA classification due to limited public S3 and DynamoDB specifications [6,24,25].

Table 3. Assumed Parameters for SCI Calculation with Sources

Parameter	Assumed Value	Source
P_{core}	1.48 W/core	InfoQ [18]
P_{mem}	0.392 W/GB	CCF Methodology [19]
P_{storage}	0.0012 W/GB	SESAME '25 [6]
$\varepsilon_{\text{network}}$	0.001 kWh/GB	SESAME '25 [6]
P_{API}	95 W	InfoQ [18]
P_{Active}	18.5 W	SESAME '25 [6] AWS EBS [24], SNIA [25]
$P_{\text{S3, active}}$	18.5 W	SESAME '25 [6] AWS EBS [24], SNIA [25]
$P_{\text{S3, idle}}$	18.5 W	SESAME '25 [6] AWS EBS [24], SNIA [25]
C_{mem}	5120 GB	AWS Specs [20]
N_{inv}	4	Estimated number of invocations per hour
S_{upload}	32666 bytes	Estimated size of a single image uploaded
PUE	1.15	AWS Report [21]
Grid EF	638 gCO_2e/kWh	ElectricityMaps [22]
Rekog. energy	70 W	NVIDIA AWS T4G DataSheet [23]

Lambda Energy Consumption

`Lambda × Compute Energy Consumption`

$$E_{L_{compute}} = \frac{1920 \times 1.48 \times 128}{1000 \times 3600 \times 1000 \times 1769} = 5.711 \times 10^{-8}\,\text{kWh} \tag{16}$$

`Lambda × Memory Energy Consumption`

$$E_{L_{memory}} = \frac{98 \times 0.392 \times (878.57 + 566)}{1024 \times 1000 \times 1000 \times 3600} = 1.50 \times 10^{-8}\,\text{kWh} \tag{17}$$

`Lambda × Storage Energy Consumption`

$$E_{L_{storage}} = \frac{12132352 \times 0.0012 \times (878.57 + 566)}{1024^3 \times 1000 \times 1000 \times 3600} = 5.44 \times 10^{-12}\,\text{kWh} \tag{18}$$

`Lambda × Network Emissions`

$$E_{\text{network}} = \frac{38804 \times 0.001}{1024^3} = 3.61 \times 10^{-8}\,\text{kWh} \tag{19}$$

Total Lambda Energy Consumption

$$E_{Lambda} = (5.711 \times 10^{-8}) + (1.50 \times 10^{-8}) + (5.44 \times 10^{-12}) + (3.61 \times 10^{-8}) \tag{20}$$

$$E_{Lambda} = 1.08 \times 10^{-7}\,\text{kWh} \tag{21}$$

S3 Energy Consumption

`S3 Active Energy Consumption`

$$E_{S3_{active}} = \frac{1 \times 13.867 \times 18.5}{1000 \times 3600 \times 1000} = 7.12 \times 10^{-8}\,\text{kWh} \tag{22}$$

`S3 Idle Energy Consumption`

$$E_{S3_{idle}} = \left(\frac{1733935 + (4 \times 32666)}{1024^3}\right) \times \left(\frac{18.5}{5120 \times 1000}\right) \times 1 \tag{23}$$

$$E_{S3_{idle}} = 6.27 \times 10^{-9}\,\text{kWh} \tag{24}$$

`Total S3 Energy Consumption`

$$E_{Storage} = E_{S3_{Active}} + E_{S3_{Idle}} \tag{25}$$

$$E_{Storage} = 7.12 \times 10^{-8} + 6.27 \times 10^{-9} = 7.74 \times 10^{-8}\,\text{kWh} \tag{26}$$

DynamoDB Energy Consumption

$$E_{\text{DDB}} = \frac{24.7527 \times 18.5}{1000 \times 3600 \times 1000} = 1.27 \times 10^{-7}\,\text{kWh} \tag{27}$$

API Gateway Energy Consumption

$$E_{\text{API Gateway}} = \frac{1878 \times 95}{1000 \times 3600 \times 1000} = 4.95 \times 10^{-5}\,\text{kWh} \tag{28}$$

Rekognition Service Energy Consumption

$$E_{\text{Rekog}} = \frac{132.91 \times 70}{1000 \times 3600 \times 1000} = 2.58 \times 10^{-6}\,\text{kWh} \tag{29}$$

Total Energy Consumption

$$E_{\text{Total}} = (1.08 \times 10^{-7}) + (7.747 \times 10^{-8}) + (1.27 \times 10^{-7}) + (4.95 \times 10^{-5}) + (2.58 \times 10^{-6}) \tag{30}$$

$$E_{\text{Total}} = 5.23 \times 10^{-5}\,\text{kWh} \tag{31}$$

Final Carbon Emission

$$SCI = 5.23 \times 10^{-5} \times 1.135 \times 638 = 0.0378\ \text{gCO}_2\text{e} \tag{32}$$

Remarks: These values are illustrative. Results depend strongly on assumed factors (P_{core}, e_{net}, e_{active}, I, etc.). As shown in Sect. 6.1, reasonable variations in these assumptions can give only an estimation of SCI. In this case, the computed value of 0.0378 gCO_2e indicates a comparatively higher carbon intensity caused majorly due to the assumed parameters and regional carbon intensity. The score could vary based on the region in which the application runs; a region with cleaner electricity source could contribute less emissions to the environment. Some other significant causes could be the use of multiple serverless functions and Visual Recognition resources. Nevertheless, the calculation demonstrates how measured metrics (time, memory, bytes) can be combined with intensity factors to yield transparent estimates.

8.3 Dashboard and Visualization Outcomes

To complement the numerical calculations, we developed CloudWatch dashboards to visualize resource usage across the workflow as well as the final carbon emission over a fixed duration. While this approach helps developers and architects view some numbers on the screen, they still cannot derive actionable insights from the dashboard since half of the values are coming from assumptions. This demonstrates a practical step towards sustainability-dashboards even within the constraints of serverless abstraction.

The following figures represent a detailed view of the energy and emission characteristics of the evaluated serverless workflow. Figure 2 shows the resource-wise energy consumption highlighting the relative contributions of individual components such as Lambda, S3, DynamoDB, API Gateway and the Rekognition Service in the workflow. Figure 3 summarizes the overall energy consumption of the workflow along with the corresponding carbon emissions providing a holistic view of sustainability impact.

Resource Wise Energy Consumption

Time	Lambda	S3	DynamoDB	API Gateway	Rekognition
Average	3.000000526507908	7.411641062020933e-8	5.047885322222222e-8	0.0001677013888888889	0.00000258432191111111

Fig. 2. Resource Wise Energy Consumption in kWh

Fig. 3. Total Energy Consumption and Final Carbon Emissions in gCO_2e

9 Implications and Conclusion

9.1 Summary of Findings

Our study demonstrates that for developers looking to design and build green software, serverless architecture may not be the best choice. Nearly half of the parameters needed for accurate calculation of SCI are unavailable due to lack of transparency. As a result, measured metrics need to be combined with assumed constants to give an emission value but this result has a wider error margin. This negative result establishes clear limits on what can and cannot be measured in serverless environments.

9.2 Practical and Research Implications

For practitioners, these findings highlight that sustainability metrics obtained in this serverless settings should be treated as approximate indicators rather than accurate calculations. Architects and developers currently cannot optimize applications based on the feedback obtained from here since critical parameters are hidden. Due to this uncertainty, development of methods for estimating hidden power-related variables is an important next step.

10 Future Work

The experimental scope of this study was limited to one small workflow, evaluated on a single cloud provider (AWS), in one region, over a short measurement window. Future work should broaden this scope to examine diverse workflow types, multiple providers and regions, and longer time spans to capture temporal and geographical variation.

While this study focuses on serverless workflows, we plan to extend this methodology to other architectures such as server-managed deployments and cloud-based applications, where the resource allocation and abstraction levels differ greatly from the serverless model. Such an extension would widen the applicability of our approach and perhaps provide a comprehensive understanding of the trade-offs among different deployment styles in terms of carbon emissions.

Furthermore, machine-learning and statistical estimation techniques could be explored to infer hidden power ratings or hardware-specific characteristics from runtime telemetry. Such techniques may reduce reliance on approximate indicators and help narrow the gap between estimated and actual energy consumption.

References

1. Naumann, S., Dick, M., Kern, E., Johann, T.: The GREENSOFT Model: a reference model for green and sustainable software and its engineering. https://www.sciencedirect.com/science/article/abs/pii/S2210537911000473. Accessed 05 Sept 2025
2. Green Software Foundation: Software Carbon Intensity (SCI) Specification. https://sci.greensoftware.foundation/. Accessed 05 Sept 2025
3. ICT Sector Guidance built on the GHG Protocol Product Life Cycle Accounting and Reporting Standard (2017). https://ghgprotocol.org/sites/default/files/2023-03/GHGP-ICTSG%20-%20ALL%20Chapters.pdf. Accessed Sept 2025
4. ISO/IEC 21031:2024 – Information technology — Software Carbon Intensity (SCI). International Organization for Standardization (2024). https://www.iso.org/standard/86612.html. Accessed Nov 2025
5. Cloud Carbon Footprint: Methodology Documentation. https://www.cloudcarbonfootprint.org/docs/methodology. Accessed 05 Sept 2025
6. Awwad, H., Lin, C., Ward, R., Shahrad, M.: Estimating the carbon footprint of serverless functions on a public cloud platform. In: SESAME 2025. ACM (2025). https://doi.org/10.1145/3721465.3721863. Accessed Nov 2025
7. Amazon Web Services: Sustainability Pillar – Serverless Applications Lens, part of the AWS Well-Architected Framework. https://docs.aws.amazon.com/wellarchitected/latest/serverless-applications-lens/sustainability.html. Accessed 05 Sept 2025
8. Amazon Web Services: AWS Serverless Developer Guide. https://docs.aws.amazon.com/serverless/. Accessed 05 Sept 2025
9. Amazon Web Services: AWS Lambda Developer Guide. https://docs.aws.amazon.com/lambda/. Accessed 05 Sept 2025
10. Amazon Web Services: Amazon S3 Documentation. https://docs.aws.amazon.com/s3/. Accessed 05 Sept 2025

11. Amazon Web Services: Amazon DynamoDB Documentation. https://docs.aws.amazon.com/dynamodb/. Accessed 05 Sept 2025
12. Amazon Web Services: Amazon API Gateway Documentation. https://docs.aws.amazon.com/apigateway/. Accessed 05 Sept 2025
13. Amazon Web Services: Amazon Rekognition Documentation. https://docs.aws.amazon.com/rekognition/. Accessed 05 Sept 2025
14. Amazon Web Services: Amazon CloudWatch User Guide. https://docs.aws.amazon.com/cloudwatch/. Accessed 05 Sept 2025
15. Amazon Web Services: Lambda Insights https://docs.aws.amazon.com/AmazonCloudWatch/latest/monitoring/Lambda-Insights.html. Accessed 05 Sept 2025
16. Amazon Web Services: Customer Carbon Footprint Tool https://aws.amazon.com/blogs/aws/new-customer-carbon-footprint-tool/. Accessed 05 Sept 2025
17. Regional Carbon Intensity. https://app.electricitymaps.com/map/72h/hourly. Accessed 05 Sept 2025
18. InfoQ: Graviton2 Benchmarks (2020). https://www.infoq.com/news/2020/03/Graviton2-benchmarks/. Accessed Nov 2025
19. Cloud Carbon Footprint. Methodology. https://www.cloudcarbonfootprint.org/docs/methodology/#memory. Accessed Nov 2025
20. Amazon S3 Documentation Specifications. https://aws.amazon.com/s3/faqs/. Accessed Nov 2025
21. AWS: Sustainability Data Centers Report (2024). https://aws.amazon.com/sustainability/data-centers/. Accessed Nov 2025
22. ElectricityMaps: Hourly Carbon Intensity Map. https://app.electricitymaps.com/map/72h/hourly. Accessed Nov 2025
23. NVIDIA: AWS T4G DataSheet (2022). https://d1.awsstatic.com/product-marketing/ec2/NVIDIA_AWS_T4G_DataSheet_FINAL_02_17_2022.pdf. Accessed Nov 2025
24. Amazon EBS volumes and NVMe. https://docs.aws.amazon.com/ebs/latest/userguide/nvme-ebs-volumes.html. Accessed Nov 2025
25. NVMe SSD Classification. https://www.snia.org/sites/default/files/SSSI/snia-nvme-ssd-classification-white-paper.pdf. Accessed Nov 2025

A Hybrid Genetic Algorithm with Restricted Mutation (GA-RM) for Task Scheduling in Cloud Computing

Gudimetla Lakshmidhara Chandan, Abhinav Kumar Singh, K. Bala Trishank(✉), and Vaidehi Vijayakumar

Vellore Institute of Technology, Chennai, India
{lakshmidhara.chandan2023,abhinavkumar.singh2023, kopparapu.bala2023}@vitstudent.ac.in, vaidehi.vijayakumar@vit.ac.in

Abstract. Efficient task scheduling is a key challenge in cloud computing, as it directly impacts resource utilization and overall execution time. This paper introduces a hybrid scheduling algorithm that integrates the exploration strength of Genetic Algorithms (GA) with a greedy allocation strategy, further enhanced through a restricted mutation mechanism. Unlike conventional GA-based methods, the proposed approach preserves the heaviest tasks while intelligently reallocating smaller tasks to reduce the maximum load across virtual machines (VMs). By protecting promising solutions from disruptive mutations, the algorithm achieves faster convergence and improved load balancing. Experimental results with independent tasks on identical VMs demonstrate that the proposed GA with Restricted Mutation (GA-RM) achieves lower makespan, higher resource utilization, and more balanced task distribution when compared with standard GA, Min-Min, Particle Swarm Optimization (PSO), and Ant Colony Optimization (ACO). The results show 2.34% to 5% improvement in makespan and 2.5% to 5.3% better resource utilization compared to traditional heuristic and metaheuristic algorithms across cloud environments.

Keywords: Task Scheduling · Genetic Algorithm · Restricted Mutation · Cloud Computing · Resource Optimization · Makespan Minimization · Hybrid Scheduling

1 Introduction

Cloud computing has transformed computing paradigms in recent years by offering flexible and instant access to storage, computing power, and services via the internet. This major change allows resources to be provisioned dynamically and services to be delivered at a low cost, thus, it is widely used in different domains such as social media, e-commerce, and Internet of Things (IoT) applications. However, the increase in the volume and heterogeneity of tasks in cloud environments has led to a demand for more efficient task scheduling mechanisms that can ensure optimal resource utilization, reduced execution time, and minimized operational costs.

© The Author(s), under exclusive license to Springer Nature Switzerland AG 2026
A. Kannan et al. (Eds.): ADCOM 2025, CCIS 2947, pp. 23–35, 2026.
https://doi.org/10.1007/978-3-032-26269-1_2

Task scheduling in cloud computing is the process of assigning computational tasks to virtual machines (VMs) in such a way as to meet various performance constraints such as makespan, throughput, and load balancing. Since the problem is NP-hard, traditional deterministic approaches do not scale efficiently and cannot dynamically adapt. As a result, heuristic and metaheuristic algorithm solutions have been considered for their effectiveness and flexibility.

One of these is Genetic Algorithms (GAs) which have become a powerful tool for scheduling due to their global search capability, resistance to local minima, and ability to adapt to dynamic workloads. GAs simulate the evolutionary process in nature through the operations of selection, crossover, and mutation, thus, producing better solutions over generations. In comparison with other metaheuristics like Particle Swarm Optimization (PSO) and Ant Colony Optimization (ACO), GAs have better mechanisms for diversity preservation and are very efficient in discrete optimization scenarios such as task-to-VM mappings. PSO is generally better suited for continuous domains and ACO for path-finding problems, GA's offer greater freedom in solution representation and operator design for scheduling applications.

However, traditional GAs are frequently plagued with problems such as premature convergence and performance degradation resulting from mutation operations being carried out in an uncontrolled manner and lack of task-specific guidance. Furthermore, current solutions rarely consider the integration of deterministic heuristics that can help rapidly reach optimal solutions and also can guide the search intelligently.

1.1 Research Gaps

A detailed review of existing literature highlights several limitations in earlier GA-based scheduling approaches:

- Unrestricted mutation often disrupts promising solutions, leading to instability and slower convergence.
- Lack of integration with greedy or heuristic techniques limits the ability to generate high-quality initial populations or perform informed mutations.
- Few studies apply adaptive or restricted mutation mechanisms tailored for cloud-specific scheduling constraints.
- Limited comparative studies benchmark hybrid GA methods against recent AI-based approaches, such as reinforcement learning or hybrid metaheuristics.

1.2 Contributions of This Work

To address these gaps, this article puts forward a Hybrid Genetic Algorithm with Restricted Mutation (GA-RM) specifically designed for scheduling tasks in heterogeneous cloud environments. The main contributions are:

- Fixing the highest-runtime task to a specific virtual machine to guide initial allocation and reduce search space complexity.
- Heuristic greedy partitioning of remaining tasks to balance runtime distribution and accelerate convergence.

- Restricted mutation strategy that limits changes to non-critical genes, preserving high-quality structures and improving stability.

This paper is organized as follows: Sect. 2 reviews existing Genetic Algorithm (GA)-based and hybrid scheduling approaches. Section 3 outlines the elements of the suggested GA-RM to include encoding, selection, crossover, and the limited mutation operator. Section 4 shows the general structure and the working process of GA-RM along with the greedy adjustment strategies. Section 5 is about the experimental findings and the comparative discussion with Min-Min, PSO, and ACO. Section 6 concludes the paper and mentions the ongoing work. Section 7 elaborates on the possible topics of research that can be explored further in this field. Lastly, Sect. 8 offers the list of references that have been cited in the paper.

2 Literature Survey

Genetic Algorithms (GAs) are well-established, effective, and highly regarded meta-heuristics for solving complex optimization problems, such as cloud task scheduling. GAs can efficiently navigate through huge solution spaces and can be adjusted to use a wide variety of different parameters. Due to these characteristics, extensive research has been conducted on GAs. In the last several years, many advancements in GA-based scheduling have been suggested. Dynamic crossover and mutation control are the product of flexible genetic operators as proposed by Ajmal et al. [1]. This type of operator improves the adaptability of Genetic Algorithms (GAs) in dynamic environments (cloud computing). Integrating secure prioritization with processor selection is how Hai et al. [2] addressed the question of how workloads can be effectively allocated across heterogeneous computing resources while maintaining security. Omara and Hamad [3] demonstrated that classical GAs are effective for scheduling independent tasks, but these techniques do not have robust solutions when managing dynamic workloads. Time Optimization Techniques and Enhanced Load Distribution were included to enhance the performance of GAs by Basahel and Yamin [4]. Their models can allocate processors to different tasks according to the time required and the current load on the system. Hybrid Heuristic Models provide an improved resource allocation strategy over several loads according to the workload, as proposed by Reddy et al. [7]. Fang et al. [8] have developed an Adaptive Mutation Mechanism that enables real-time adjustment of the mutation frequency in order to help the GA converge more quickly to the optimal solution. Ibrahim et al. [9] demonstrated how to build a Cost-Aware Scheduling Model which provides cost-efficient scheduling by jointly minimizing makespan and total cost. High-Performance Task Scheduling was also improved with Adaptive Crossover Techniques, as explained by Singh and Kalra [10].

Over the past few years, researchers have been increasingly interested in incorporating Genetic Algorithms (GA) with Artificial Intelligence (AI) and other learning methodologies. Recently Shao et al. [11] proposed a novel hybrid method called PGA which integrates GA with Particle Swarm Optimization (PSO) to provide an improved method for scheduling task deadlines on distributed systems and achieving superior convergence speeds and solution quality compared to the traditional standalone techniques. Xiu et al. [12] devised an adaptive scheduling system based on Meta Reinforcement

Learning (MRL), to optimize task allocation based upon the different types of resources present within Cloud Computing but providing improved flexibility and rapid responses to assigned tasks. Abdel-Basset et al. [13], introduced two types of metaheuristic models to allocate resource tasks based upon energy-efficient allocation schemes to achieve a significant decrease in fog-based Internet of Things (IoT) platforms' energy consumption while still meeting Watson's (Quality of Services) criteria. Ding et al. [14] performed dynamic scheduling of Cloud Computing tasks using the Q-learning methodology to provide maximum energy efficiency and real-time updates on optimal scheduling policies. Kumar & Bhagwan [15] performed a comparison of several meta-heuristic scheduling algorithms such as GA, ACO and PSO and found hybrid methods to be more efficient than their pure counterparts with regards to execution time and scalability while Kaur & Mann [16] introduced a hybrid genetic-firefly algorithm which reduces the costs of executing workflows as well as enhancing the efficiency of cloud-based workflow scheduling.

However, there are still some issues that have not been fully addressed by these developments. In numerous GA variants, mutations are often poorly controlled, which disrupts high-quality solutions. Few works have considered the use of deterministic greedy strategies for initial population generation or constraint enforcement. Besides that, few works have considered the use of adaptive and restricted mutation operators, despite their potential to speed up convergence and keep the solution unchanged. Most of the research works conducted have provided limited comparative benchmarking with AI-driven schedulers.

To address this limitation, a Hybrid Genetic Algorithm with Restricted Mutation (GA-RM) is introduced in this article. The new approach fixes the longest task with one of the available virtual machines, thus making it easier and more stable to take further scheduling decisions. The rest of the tasks are distributed by a greedy partitioning method aimed at balancing the workload. Mutation operations are selectively restricted to specific chromosome positions, preserving high-quality partial solutions while reducing disruptive variations. By combining heuristic allocation with evolutionary search, the GA-RM approach improves both the efficiency and robustness of GA-based schedulers, particularly in heterogeneous cloud environments.

3 Description of Hybrid Genetic Algorithm

This section describes the main building blocks of the proposed Hybrid Genetic Algorithm with Restricted Mutation (GA-RM). The presentation is organized into the following components: encoding scheme, selection, crossover, mutation, fitness evaluation, and the motivation for incorporating greedy heuristics together with mutation restrictions.

3.1 Encoding Strategy

Tasks	T1	T2	T3	T4	T5	T6	T7
Virtual Machine	VM2	VM1	VM5	VM4	VM3	VM4	VM5
Runtime	t_0	t_1	t_6	t_5	t_4	t_3	t_2

Fig. 1. Gene Representation

In GA-RM, a chromosome represents the allocation of tasks to VMs. If there are m tasks and n VMs, each gene in a chromosome denotes the VM assigned to a particular task.

3.2 Selection Mechanism

In GA-RM, we use a roulette wheel selection method to choose solutions for the next generation. Because our goal is to minimize makespan and cost, the selection is based on the inverse of the fitness value, so that better (lower-fitness) solutions have a higher chance of being picked. At the same time, weaker solutions are not completely excluded, which helps maintain diversity in the population. This balance between favoring good solutions and keeping variety prevents premature convergence and makes the search process more reliable.

3.3 Crossover Operation

In GA-RM, we use a uniform crossover method to create new solutions from two parent chromosomes. At each gene position, a random choice is made to decide whether the task assignment comes from the first or the second parent. This way, the offspring inherit a well-mixed combination of genes, instead of taking a single block from one parent as in single-point crossover. The fine-grained mixing helps maintain diversity in the population and increases the chances of finding better scheduling solutions.

3.4 Restricted Mutation Operator

In the proposed GA-RM framework, mutation is applied in a restricted manner to preserve high-quality gene sequences and prevent disruption of critical task assignments. Specifically, the mutation operator identifies the heaviest task, i.e., the task with the maximum runtime, and fixes its allocation to the initially assigned VM. The remaining tasks are then reallocated using a greedy strategy that assigns each task to the VM with the current minimum load. This selective mutation mechanism ensures that the heaviest task remains unaffected, thereby preserving the structural integrity of the solution, while allowing flexibility in redistributing the other tasks to improve load balancing. By reducing randomness in the assignment of high-runtime tasks, the restricted mutation accelerates convergence and enhances the stability of the evolutionary search process.

3.5 Fitness Function

The fitness function of the proposed method aims to minimize the makespan which is the maximum completion time among all virtual machines (VMs). Suppose T = {t1, t2,... tn} is the set of tasks, where each task ti has an execution time and V = {v1, v2, v3,... vm} is the set of m VMs. One gene can represent the matching of tasks to VMs. The load on a VM Vj can be

$$L_j = \sum_{i=1}^{n} \mathrm{t_i} \cdot \delta\left(A(t_i), v_j\right) \tag{1}$$

where $\delta\left(A(t_i), v_j\right) = 1$ if task i is assigned to VM j, and 0 otherwise. The fitness function is then defined as

$$f(A) = max_{\{1 \leq j \leq m\}} L_j \tag{2}$$

representing the makespan of a given allocation. A reduction in this function not only leads to a balanced workload distribution but also decreases the total completion time and has a positive effect on the resource utilization as well as the system throughput.

3.6 Heuristic Greedy Partitioning

To improve performance of the initialization and mutation phases a guided task allocation strategy is added to the GA-RM framework. The method begins by identifying the task with the longest execution duration and assigning it to a specific virtual machine (VM). Thus, the heaviest workload remains consistent across generations. After that, the rest of the tasks are taken up according to their runtimes and are sequentially assigned to the VM which currently has the lowest load. With each assignment, the load of the VM is updated and this operation goes on until all tasks have been assigned. This technique ensures balanced task distribution; it keeps the stability of the vital resource allocation. By fixing the most resource-demanding task and utilizing a systematic allocation procedure for the rest, the algorithm achieves better load balance, less makespan, and faster convergence as compared to usual random methods. Overall, this strategy helps the algorithm to avoid weak initial configurations and decreases the number of unnecessary adjustments during the evolutionary process.

4 Proposed Hybrid GA-RM

4.1 Architecture

The proposed method features a hybrid Genetic Algorithm integrated with a restricted mutation strategy to effectively resolve a task scheduling problem in a heterogeneous cloud computing environment. The fundamental design has been separated into three major stages: population initialization, evolution through selection, crossover, and mutation, and fitness-based scheduling. Thus, each task is assigned to a VM such that the total completion time, i.e., makespan, would be minimal, and load would be balanced across all VMs. The scheduling is done through a heuristic-based initialization that starts off by assigning the task with the highest runtime to the first VM. This, in turn, serves as a reference point for balancing the workload in other VMs. The rest of the tasks are allocated in such a way that the total load of each VM is as close to the maximum runtime as possible without crossing it.

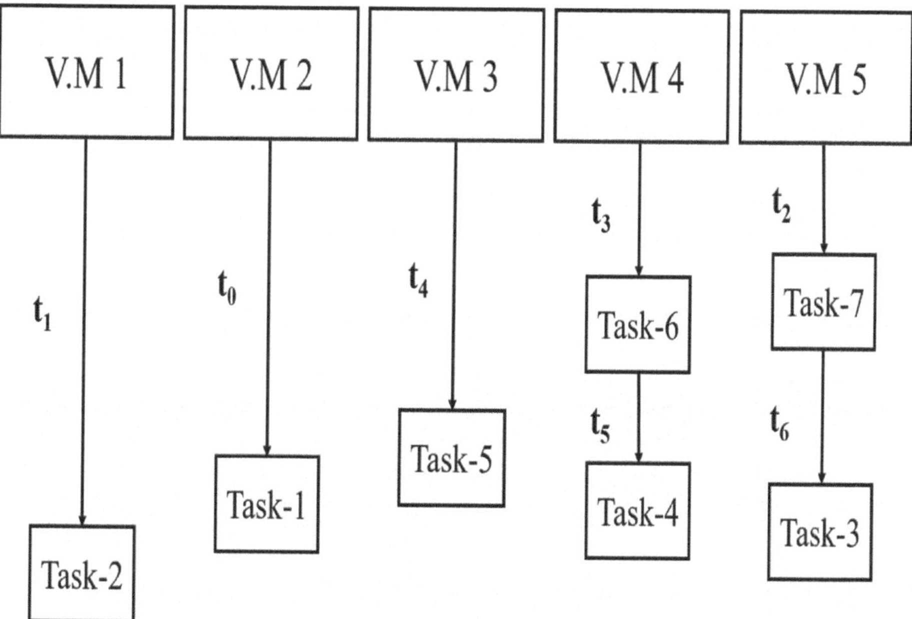

Fig. 2. Architecture of Virtual Machine Allocation

Virtual Machine	V2	V1	V5	V4	V3	V4	V5
Task Runtime	t_0	t_1	t_2	t_3	t_4	t_5	t_6

Fig. 3. Runtime of each task assigned to its respective virtual machine

4.2 Restricted Mutation and Greedy Adjustment

To enhance the quality of the solution and avoid the premature convergence, a limited mutation mechanism is introduced. In this mechanism:

- The tasks with the maximum runtimes at each iteration are not subjected to mutation in order to keep the balance that was already established during the initial allocation.
- Mutation is applied only to the remaining tasks, allowing small adjustments while preserving the integrity of critical assignments.
- If the algorithm is unable to discover a better solution after a certain number of generations; it forcibly mutates to bring back diversity and get out of local optima.

This greedy-mutation hybrid enables faster convergence by less flexible reallocation of smaller ones while more flexible anchoring of high-load tasks.

The proposed algorithm may also be suitable for situations with interdependent tasks. In that case, task dependencies are maintained by limiting the interdependent tasks that have to be executed on the same virtual machine especially when their total runtime is considerable. Thus, communication overhead is minimized and scheduling performance remains efficient.

4.3 Algorithm

The GA-RM algorithm follows these main steps:

1. Initialize population of chromosomes
   → Each chromosome represents a task-to-VM mapping
   → Use greedy heuristic to assign tasks in descending order of run time to VMs with least current load.

2. Fitness of each chromosome is evaluated where:
   → Fitness = maximum makespan load across all VMs

3. While TRUE:
   → Select parent chromosomes using roulette wheel selection.
   - Randomly select two chromosomes
   - Choose the one with better fitness for reproduction

   → Apply uniform crossover to generate offspring

   → **Apply restricted mutation to offspring**
   - **Protect top-t longest tasks from mutation.**
   - **Reassign other task mappings with a given mutation probability.**
   - **New VM is selected randomly from available VMs.**

   → **heuristic_allocation(chromosome, task_times, num_vms)**
   - **Sort tasks by runtime in descending order**
   - **Initialize vm_loads ← [0] * num_vms**
   - **Initialize allocation ← [] for _ in range(num_vms)**
   - **For each (task_index, task_time) in sorted list:**
     - **min_vm ← VM with minimum load.**

   → Assign task_index to min_vm, update load
   - For each VM:
     - Update chromosome[task_index] ← vm for all tasks assigned.
     - Return updated chromosome.

   → Evaluate fitness of new offspring population.
   → Apply elitism to preserve best chromosome in next generation.

4. Update global best solution if improved.

5. Return the best chromosome with minimum makespan.

5 Results and Discussion

The experiments were run on 500 computational jobs that were scheduled over 20 virtual machines (VMs). The new Hybrid Genetic Algorithm with Restricted Mutation (GA-RM) was compared with three conventional task scheduling methods: Min-Min,

Particle Swarm Optimization (PSO), and Ant Colony Optimization (ACO). The main performance metrics considered were makespan, resource utilization, and throughput.

According to the makespan comparison (Fig. 4), GA-RM has the lowest makespan, which means that it can schedule more efficiently than other algorithms. In fact, GA-RM is 2.34% better than Min-Min, 19.49% better than PSO, and 5.02% better than ACO.

The utilization indicator (Fig. 5) also reveals the effectiveness of GA-RM as it almost reached 99.89% while Min-Min, PSO, and ACO were only able to achieve 97.56%, 80.41%, and 94.88%, respectively. Thus, VM resources are better utilized with less vacant time. The throughput figures (Fig. 6) are in line with these findings and demonstrate that GA-RM is the most productive in terms of the task execution rate by achieving 0.0679 tasks/unit time, which is just a bit higher than that of Min-Min and ACO and significantly higher than that of PSO.

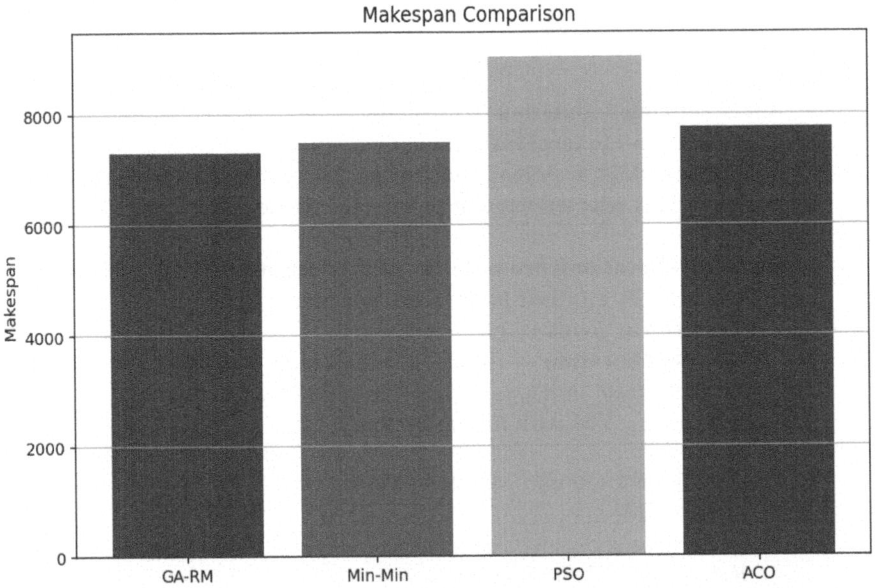

Fig. 4. Makespan Comparison

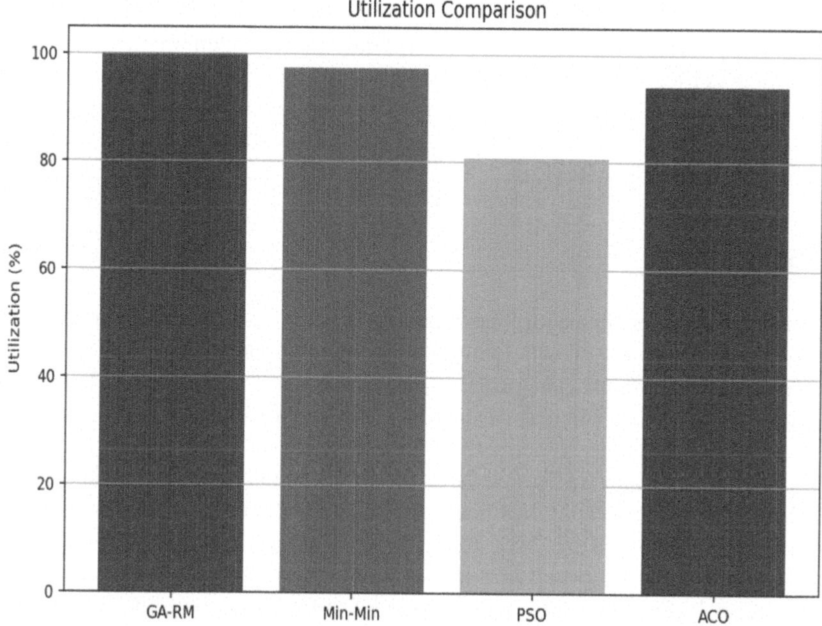

Fig. 5. Utilization Comparison

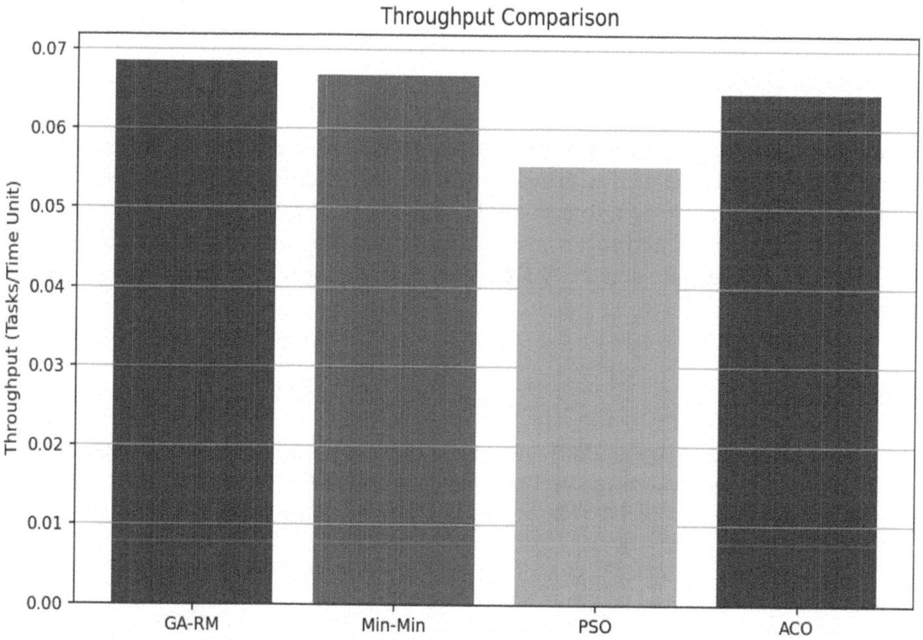

Fig. 6. Throughput Comparison

In general, the GA-RM algorithm consistently performs equal to or better than other algorithms in all its dimensions of evaluation. These findings demonstrate the proposed hybrid method's efficiency and particularly in heterogeneous cloud environments, where reducing makespan and improving resource utilization remain critical challenges.

The source code of the proposed algorithm is publicly available on GitHub at https://github.com/gl-chandan/GA-RM.git.

6 Conclusion

A Hybrid Genetic Algorithm with Restricted Mutation (GA-RM) was the main focus of this study. Its objective was to efficiently handle the task scheduling problem in heterogeneous cloud environments. The proposed method, by fixing maximum-runtime tasks, using greedy partitioning for rest of the tasks, and restricting mutation operations to non-critical genes, manages to achieve faster convergence, better stability, and higher scheduling efficiency than traditional GAs and other metaheuristics. The experimental results show that the makespan has been improved by 2.34% to 5% and resource utilization has been increased by 2.5% to 5.3%, as well as a considerable enhancement in throughput, thus the proposed GA-RM model is validated as an effective one. The implementation, based on Python and machine-learning-based experiments, also proves its versatility and reproducibility in real-life situations. Future work may consider combining reinforcement learning for energy-aware scheduling and large-scale datasets to improve performance under varied and changing cloud workloads. The team is currently working on the performance of the proposed GA-RM scheme in heterogeneous virtual machine environments and scheduling scenarios with task dependencies.

7 Future Work

We are examining the performance of the proposed GA-RM scheme for heterogeneous virtual machines and scenarios with task dependencies. Additionally, future research will focus on extending the system to account for Quality of Service (QoS) features such as deadlines, task priorities, and cost considerations. These upgrades intend to enable multi-objective optimization and make GA-RM more feasible for real-world cloud computing environments.

References

1. Ajmal, M.S., Iqbal, Z., Umair, M.B., Arif, M.S.: Flexible genetic algorithm operators for task scheduling in cloud datacenters. In: 2020 14th International Conference on Open Source Systems and Technologies (ICOSST), pp. 1–6. Institute of Electrical and Electronics Engineers Press (2020). https://doi.org/10.1109/ICOSST51357.2020.9333057
2. Hai, T., Zhou, J., Jawawi, D., et al.: Task scheduling in cloud environment: optimization, security prioritization and processor selection schemes. J. Cloud Comput. **12**, 15 (2023)
3. Hamad, S.A., Omara, F.A.: Genetic-based task scheduling algorithm in cloud computing environment. Int. J. Adv. Comput. Sci. Appl. **7**(4) (2016)

4. Basahel, S.B., Yamin, M.: A novel genetic algorithm for efficient task scheduling in cloud environment. In: 2022 9th International Conference on Computing for Sustainable Global Development (INDIACom), pp. 30–34. IEEE Press (2022). https://doi.org/10.23919/INDIACom54597.2022.9763230
5. Khaledian, N., Voelp, M., Azizi, S., et al.: AI-based and heuristic workflow scheduling in cloud and fog computing: a systematic review. Cluster Comput. **27**, 10265–10298 (2024)
6. Lambora, A., Gupta, K., Chopra, K.: Genetic algorithm: a literature review. In: 2019 International Conference on Machine Learning, Big Data, Cloud and Parallel Computing (COMITCon), pp. 380–384. IEEE Press (2019). https://doi.org/10.1109/COMITCon.2019.8
7. Reddy, P.S., Trisha, K., Sree, N.H., Bhaskaran, S., Sampath, N.: Enhanced task scheduling in cloud computing: a comparative analysis and hybrid algorithm implementation using CloudSim. In: 2024 15th International Conference on Computing Communication and Networking Technologies (ICCCNT), pp. 1–5. IEEE Press (2024). https://doi.org/10.1109/ICCCNT61001.2024.10725185
8. Yiqiu, F., Xia, X., Junwei, G.: Cloud computing task scheduling algorithm based on improved genetic algorithm. In: 2019 IEEE 3rd Information Technology, Networking, Electronic and Automation Control Conference (ITNEC), pp. 852–856. IEEE Press (2019). https://doi.org/10.1109/ITNEC.2019.8728996
9. Ibrahim, E., El-Bahnasawy, N.A., Omara, F.A.: Task scheduling algorithm in cloud computing environment based on cloud pricing models. In: 2016 World Symposium on Computer Applications and Research (WSCAR), pp. 65–71. IEEE Press (2016). https://doi.org/10.1109/WSCAR.2016.20
10. Singh, S., Kalra, M.: Scheduling of independent tasks in cloud computing using modified genetic algorithm. In: 2014 International Conference on Computational Intelligence and Communication Networks, pp. 565–569. IEEE Press (2014). https://doi.org/10.1109/CICN.2014.128
11. Shao, K., Song, Y., Wang, B.: PGA: A new hybrid PSO and GA method for task scheduling with deadline constraints in distributed computing. Mathematics **11**, 1548 (2023). https://doi.org/10.3390/math11061548
12. Xiu, X., Li, J., Long, Y., et al.: MRLCC: an adaptive cloud task scheduling method based on meta reinforcement learning. J. Cloud Comput. **12**, 75 (2023). https://doi.org/10.1186/s13677-023-00440-8
13. Abdel-Basset, M., El-Shahat, D., Elhoseny, M., Song, H.: Energy-aware metaheuristic algorithm for industrial-internet-of-things task scheduling problems in fog computing applications. IEEE Internet Things J. (2020). https://doi.org/10.1109/JIOT.2020.3012617
14. Ding, D., Fan, X., Zhao, Y., Kang, K., Yin, Q., Zeng, J.: Q-learning based dynamic task scheduling for energy-efficient cloud computing. Future Gener. Comput. Syst. **108** (2020). https://doi.org/10.1016/j.future.2020.02.018
15. Kumar, R., Bhagwan, J.: A comparative study of meta-heuristic-based task scheduling in cloud computing. In: Dubey, H.M., Pandit, M., Srivastava, L., Panigrahi, B.K. (eds.) Artificial Intelligence and Sustainable Computing. Algorithms for Intelligent Systems. Springer, Singapore (2022). https://doi.org/10.1007/978-981-16-1220-6_12
16. Kaur, I., Mann, P.S.: A hybrid cost-effective genetic and firefly algorithm for workflow scheduling in cloud. In: Gupta, D., Khanna, A., Bhattacharyya, S., Hassanien, A.E., Anand, S., Jaiswal, A. (eds.) International Conference on Innovative Computing and Communications. AISC, vol 1166. Springer, Singapore (2021). https://doi.org/10.1007/978-981-15-5148-2_4

Multi-modal Automatic Detection of Distress Behaviors in Surveillance Systems

K. Gayathri(✉), A. Chitra, and S. Priya Dharshini

PSG College of Technology, Peelamedu, Coimbatore, India
{kgi.mca,ac.mca}@psgtech.ac.in

Abstract. Uncommon behavior detection in surveillance systems is a critical component of public safety, particularly in scenarios where early intervention can mitigate potential harm. This paper presents a multi-modal deep learning framework for surveillance-driven distress detection, with a specific emphasis on ensuring individual safety and well-being. The proposed framework provides a holistic analysis of distress related scenarios by combining MobileNetV2, fine-tuned on the FER-2013 dataset for emotional recognition from the face, MediaPipe Pose for estimating the posture, and DeepFace for identifying the gender. Upon detecting the distress signs such as anger, fear or any defensive body postures, the system activates an automated alert mechanism that notifies pre-registered emergency contacts, ensuring quick intervention. The results from the experiments demonstrate the feasibility of the proposed framework for distress detection, with strong potential for real-time deployment in video surveillance and emergency response systems. By combining efficient deep learning models with behavior-driven surveillance analysis, the proposed work contributes towards the development of scalable, proactive, and socially impactful surveillance systems aimed at enhancing safety and security.

Keywords: Emotion Recognition · Human Pose Estimation · Deep Learning · Gender Classification · Automated Alert System · Surveillance systems

1 Introduction

Distress detection specializes in detecting distress signs in a person by examining facial expressions and body posture. Processing the images for the detection of distress-based emotions like fear and anger and body postures indicating an emergency like defensive stances or a posture is included. Added to this, gender prediction increases the precision of detection through contextual awareness. By integrating these analyses, distress detection systems facilitate the identification of possible threats and automated alerting, such that emergency responses are faster and safer in public and private environments.

1.1 Data Representation

The system processes them by utilizing several pre-processing methods to ensure accuracy and consistency in gender prediction, posture classification, and emotion recognition. Expressions in face are analyzed through a deep learning model trained on the

© The Author(s), under exclusive license to Springer Nature Switzerland AG 2026
A. Kannan et al. (Eds.): ADCOM 2025, CCIS 2947, pp. 36–50, 2026.
https://doi.org/10.1007/978-3-032-26269-1_3

FER-2013 dataset, with input images undergoing preprocessing steps such as fixed dimension resizing, normalization on the pixel value, and conversion of grayscale to improve feature extraction. MediaPipe Pose is used to estimate the pose by identifying the key body landmarks to generate a set of keypoints which describe the body posture. By extracting the facial features and mapping them to the predefined gender classes, gender prediction is done using DeepFace in-order to provide contextual information for distress analysis. To achieve accurate classification, the extracted features are converted into structured formats compatible with machine learning models: facial emotions are encoded as numerical vectors, pose keypoints are represented through coordinate features, and gender outputs are expressed as class-wise probability scores. This structured representation facilitates efficient processing and enhances the robustness of distress detection in real-world scenarios.

1.2 Classification Approaches

The proposed framework incorporates three distinct classification models-facial emotion recognition, body posture analysis, and gender classification-each leveraging specialized feature extraction methods and mathematical operations to evaluate distress levels. For emotion recognition, the system employs MobileNetV2, a CNN fine-tuned on the FER-2013 dataset. The model extracts spatial features from facial images using convolution and depthwise seperable convolution layers, which minimize computational overhead while retaining critical features. A Softmax activation function is used to achieve the final emotion classification that converts the model output z_i into class probabilities, ensuring they collectively sum to one. The emotion label associated with the highest probability is then selected, categorizing the facial expression into fear, sadness, anger, or neutral

$$\mathrm{p}(y_i) = \frac{e^{z_i}}{\sum_{j=1}^{n} e^{z_j}} \tag{1}$$

.

MediaPipe Pose is used to estimate the pose with 33 key body landmarks assigned to the (x, y, z) coordinates to each joint. Postures are classified by computing joint angles between selected keypoints. The system applies the cosine rule, where distances between joints such as the hip, knee, and ankle are used to calculate angles. The computed joint angles are evaluated against predefined thresholds to classify various postures. Specifically, a hip-knee-ankle angle of approximately 180° indicates a standing position, while an angle less than 120° corresponds to a sitting posture. Similarly, a shoulder-hip inclination greater than 45° is used to identify a lying-down position.

$$\theta = cos^{-1}\left(\frac{A^2 + B^2 - C^2}{2AB}\right) \tag{2}$$

DeepFace utilizes facial embeddings that are generated by pre-trained deep CNNs such as FaceNet and VGG-Face to classify the gender. These embedding encode facial features into a 128-dimensional vector representation, which is then compared with reference gender clusters using cosine similarity. Given two feature vectors, A (input face) and B (reference cluster), the similarity score determines how closely the input aligns with each gender class. The model outputs probability scores for both male and female categories, with the highest score used to predict the final gender label.

$$\cos(\theta) = \frac{A.B}{\|A\| \|B\|} \tag{3}$$

1.3 Novelty of the Framework

The proposed framework introduces several novel aspects compared to existing surveillance and safety systems:

- Multi-Modal Fusion for Distress Detection: Traditional methods rely on a single modality, whereas, this framework combines facial emotion recognition, body posture estimation and gender classification to improve robustness and contextual understanding.
- Distress-Oriented Behavior Modeling: This work introduces distress-specific cues (fear, anger, defensive postures) as indicators of uncommon behavior, targeting psychological and physical states linked to safety.
- Safety-Driven Alert Mechanism: The system does not just classify behaviors but also triggers automated notifications to emergency contacts, bridging the gap between detection and intervention.

2 Literature Review

Recent advances in deep learning and computer vision have driven significant progress in body posture analysis and emotion recognition, both essential for distress detection in safety applications. Emotion recognition interprets facial expressions, while posture analysis evaluates human poses to identify potential distress. Deep learning models have improved the accuracy and reliability of these tasks, enabling more effective real-time detection. This review highlights key studies, their methodologies, datasets, and findings, with approaches ranging from CNNs, transfer learning, and graph-based models to hybrid architectures and sensor-based tracking systems. However, challenges such as occlusions, environmental variations, and dataset imbalances still limit detection accuracy and generalizability.

Jonathan et al. (2020) [4] investigated facial emotion recognition using the FER-2013 dataset, which includes 35,887 grayscale images across seven emotion categories: Angry, Disgust, Fear, Happy, Sad, Surprise, and Neutral. They employed transfer learning by fine-tuning the VGG-16 model, a deep CNN known for its depth and feature extraction capability, achieving an accuracy of 69.40%. The study demonstrates the effectiveness of transfer learning in facial emotion classification, showing how pre-trained models can address challenges such as limited labeled data and high computational costs, while fine-tuning enhances generalization across diverse facial expressions.

Ali et al. (2022) [1] carried out a detailed review of human body posture recognition techniques, with particular emphasis on approaches like OpenPose and deep learning-based classification using VGG-16. Their evaluation across multiple datasets demonstrated the effectiveness of deep learning, yielding impressive classification accuracies of 95% for walking, 87.4% for hugging, and 90.1% for fighting. In addition to showcasing performance, the study also examined key challenges that continue to affect posture recognition systems, such as occlusions, variations in body shape, and diverse environmental conditions. The authors stressed that overcoming these limitations requires the use of large and diverse datasets, as well as robust feature extraction strategies to enhance generalization.

Expanding on these directions, Liu et al. (2022) [7] introduced an automatic posture recognition framework that integrated Kinect sensors with Graph Convolutional Networks (GCNs). This approach leveraged joints, enabling more accurate modeling of complex human postures. This system achieved a classification accuracy of 92.2% and demonstrated strong resilience against differences in body types and variations in movement patterns. Importantly, the adaptability of their method highlights its potential for practical real-world applications, particularly in fields such as healthcare monitoring, rehabilitation, and sports analysis, where accurate posture recognition is essential for tracking physical activities and ensuring safety.

Lee et al. (2022) [5] developed a deep learning-based posture recognition framework specifically designed for real-time tracking with Unmanned Aerial Vehicles (UAVs). The system was engineered to address common challenges in outdoor monitoring, such as occlusions and environmental variations, ensuring consistent performance under dynamic conditions. With a reported classification accuracy of 95.2%, the model demonstrated high reliability and precision, making it particularly well-suited for large-scale applications. Its effectiveness was highlighted in critical use cases such as surveillance, disaster management, and search-and-rescue operations, where timely and accurate posture recognition can significantly enhance situational awareness and support rapid decision-making.

Liaqat et al. (2021) [13] developed a hybrid posture detection framework that combines CNNs with LSTM networks to capture both spatial and temporal features, attaining an accuracy of 93.5% on a large dataset. Their findings highlight the value of hybrid architectures in enhancing robustness, particularly for sequential posture analysis in applications like healthcare monitoring and sports tracking.

Similarly, Tang et al. (2021) [14] introduced a wearable posture classification system using inertial sensors integrated with RNNs, where LSTM layers effectively captured dynamic upper-body posture transitions. Their system demonstrated high accuracy across seven static sitting postures, proving the potential of sensor-deep learning integration for healthcare, rehabilitation, and workplace ergonomics, with possible extension to full-body classification.

Complementing these studies, Urnisha et al. (2022) [11] focused on facial emotion recognition by applying transfer learning techniques. They fine-tuned the MobileNetV2 model on the FER-2013 dataset, which consists of diverse facial expressions across multiple categories. Their approach achieved an accuracy of 61%, reflecting the challenges inherent in emotion detection tasks compared to posture recognition, particularly due to variations in facial expressions, lighting conditions, and occlusions. Although its performance was lower compared to heavier architectures, MobileNetV2's lightweight design makes it well-suited for mobile and embedded platforms, highlighting the importance of balancing computational efficiency and recognition accuracy in real-time emotion recognition tasks.

Overall, prior studies in facial emotion recognition and human posture analysis demonstrate the effectiveness of deep learning and transfer learning approaches for behavior understanding in surveillance and safety applications. While models such as VGG-16 and MobileNetV2 have been successfully fine-tuned on datasets like FER-2013 to recognize emotions with varying levels of accuracy, other works have explored lightweight architectures for mobile deployment, highlighting trade-offs between efficiency and performance. In posture recognition, techniques ranging from OpenPose-based classification to advanced CNN-LSTM and GCN architectures, as well as wearable sensor-based systems, have achieved high accuracies across diverse scenarios such as healthcare, workplace safety, and surveillance. Furthermore, studies leveraging UAVs and hybrid models emphasize robustness in dynamic and occlusion-prone environments. This body of work establishes a strong foundation for multi-modal analysis, while also underscoring gaps in distress-specific behavior detection within surveillance systems, thereby motivating the present research.

3 Methodology

3.1 Data Collection

The FER-2013 dataset [16] serves as the primary dataset for emotion recognition. It consists of 35,887 grayscale images of facial expressions categorized into seven emotions: Happy, Sad, Angry, Fear, Surprise, Disgust, and Neutral. Since the focus of this project is distress detection, only three classes (Sad, Angry, and Fear) were selected. Sample images from the dataset under each category is given in Fig. 1. Other emotions are classified as normal to prevent unnecessary misclassifications and ensure that the system remains aligned with its primary objectives.

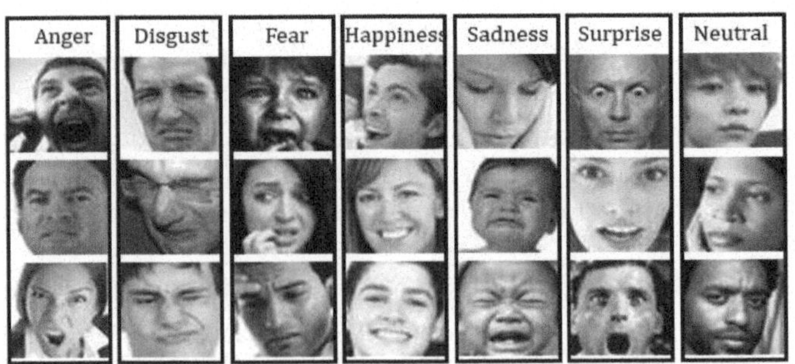

Fig. 1. Sample Images from FER-2013 dataset

3.2 Key Algorithms and Techniques

3.2.1 Deep Learning Models for Facial Emotion Recognition

The primary objective was to identify a model that balances accuracy and training efficiency on the FER-2013 dataset. A basic CNN trained from scratch achieved only 55%

accuracy, proving inadequate. VGG16, though powerful, was too slow for practical use, while ResNet20 offered strong feature extraction but suffered from long training times. Ultimately, MobileNetV2, a lightweight CNN pre-trained on ImageNet, was selected for its efficiency, faster training, and suitability for real-time applications. It achieved 66% test accuracy, outperforming the basic CNN, with early stopping applied at epoch 19 to prevent over-fitting. Its architecture, featuring inverted residuals and linear bottlenecks, ensures reduced computational complexity while maintaining competitive accuracy, making it the most practical choice for emotion recognition in this system.

3.2.2 Emotion Detection Pipeline

The emotion recognition process follows a structured pipeline to ensure accurate predictions, beginning with face detection using OpenCV's Haar cascade or DeepFace's built-in detector. The identified face regions are then resized and normalized before being passed into the MobileNetV2 model, which classifies them into one of three distress-related emotions. This streamlined process enables the system to efficiently analyze facial images and deliver reliable emotional assessments. The emotion recognition model processes an input image as a matrix $I(x,y)$ representing pixel intensities. The CNN layers apply convolutional operations using kernels K, which are defined where $F(x,y)$ represents the output feature map, $K(i,j)$ is the kernel filter, and the $I(x+i,y+j)$ corresponds to pixel intensities in the input image. To obtain class probabilities, the softmax activation function is applied.

$$F(x, y) = \sum_{i=-k}^{k} \sum_{j=-k}^{k} K(i, j).I(x + i, y + j) \tag{4}$$

3.2.3 Pose Estimation Model

Pose estimation is a critical component of the distress detection system, as body posture provides key indicators of a person's state, particularly distress-related postures such as falling, lying down, or defensive stances. Initially, OpenPose was considered due to its high accuracy and ability to detect multiple keypoints, making it effective for analyzing subtle bodily cues; however, installation complexities involving dependencies like Caffe, CUDA, and cuDNN, along with performance constraints such as slow inference on consumer-grade GPUs, made it impractical for seamless integration. To address these challenges, MediaPipe Pose was adopted as the final solution, offering a lightweight, efficient, and real-time alternative developed by Google. Unlike OpenPose's heavy reliance on deep learning models, MediaPipe [15] combines deep learning with geometric processing for faster keypoint detection, predicting 33 landmarks compared to OpenPose's 17, shown in Fig. 2. thereby enhancing posture-based distress recognition. With simple integration requirements, cross-platform support, and the ability to run efficiently on CPUs without high-end GPUs, MediaPipe ensures smooth performance across devices, making it well-suited for real-time distress detection in diverse environments.

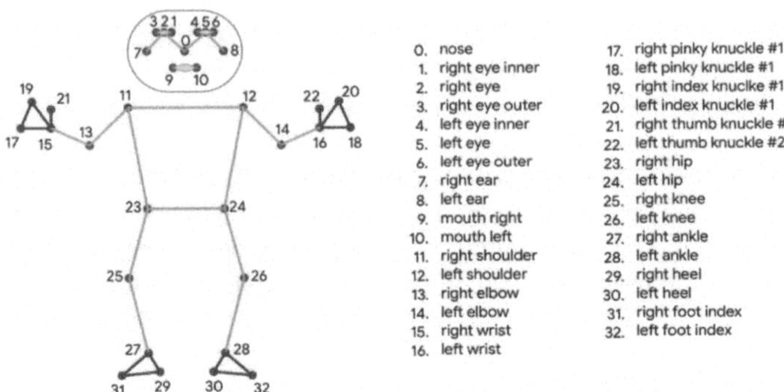

Fig. 2. Keypoints of MediaPipe pose

3.2.4 DeepFace

DeepFace is an open-source deep learning library that supports multiple state-of-the-art face recognition models such as VGG-Face, Google FaceNet, and OpenFace, and also provides built-in support for gender classification using pre-trained deep convolutional neural networks (CNNs) to classify an individual's gender as either Man or Woman. Unlike building a separate gender classification model, DeepFace offers a ready-to-use solution with high accuracy, as illustrated in Fig. 3. Showing the DeepFace representation architecture [15]. Its key advantages include being pretrained on large-scale face datasets, ensuring reliable classification across diverse ethnicities and lighting conditions; eliminating the need for additional training, dataset collection, annotation, or heavy computation; seamless integration with Flask as a REST API for real-time applications; and achieving over 95% accuracy on well-lit, frontal face images.

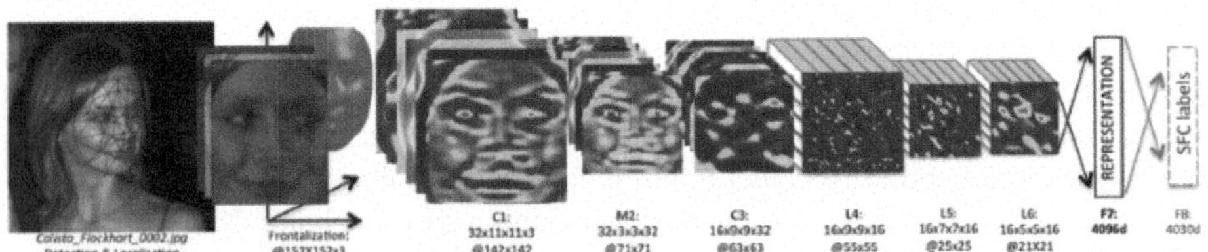

Fig. 3. DeepFace Architecture

3.3 Data Preprocessing

To ensure effective feature extraction, the system applies a series of preprocessing steps to standardize input images before deep learning analysis. Images are resized to 224x224 pixels to meet the input specifications of mobileNetV2 for emotion recognition and MediaPipe Pose for posture estimation, followed by pixel normalization to the range [0, 1] for stable inference. For facial emotion analysis, MTCNN [17] is used to detect and crop face regions, ensuring proper alignment, while MediaPipe Pose extracts 33 skeletal landmarks for posture classification. To improve model generalization and reduce overfitting, data augmentation techniques-including random rotation, brightness adjustments, and flipping are applied, thereby increasing dataset diversity. Together, these preprocessing and augmentation strategies enhance the robustness and reliability of the proposed distress detection system. The overall workflow of the system is depicted in Fig. 4, with each step detailed in the following sections.

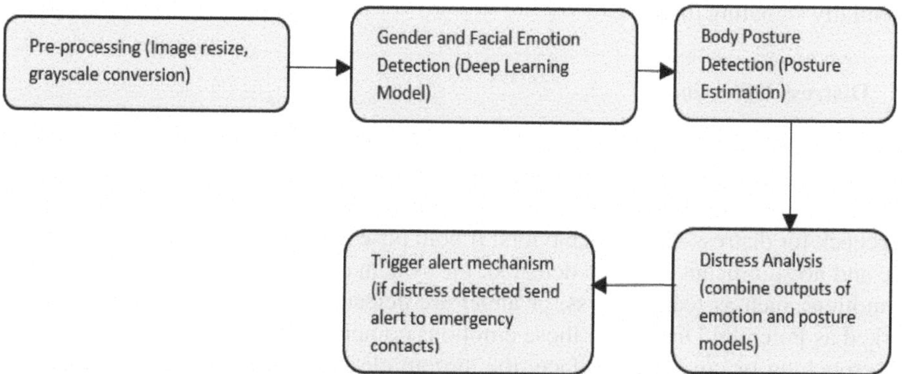

Fig. 4. Workflow of the Framework

3.4 Emotion Recognition-Feature Extraction

A fine-tuned MobileNetV2 model, trained on the FER-dataset, is employed for facial emotion recognition, where detected face regions are classified into three distress-related emotions: fear, sadness, and anger. MobileNetV2 employs depthwise separable convolutions to efficiently extract deep facial features while minimizing computational cost by generating feature maps through convolution operations, where pixel values are combined with learned weights and adjusted by bias terms. Each activation is computed with respect to a specific input channel, allowing filters to independently capture subtle variations in facial expressions. Following this, a 1×1 pointwise convolution integrates information across channels, strengthening the overall representation learning. The transformed feature maps are then passed through a fully connected layer with a softmax activation function that assigns probability scores to each emotion category, with the highest probability determining the final label, thereby ensuring reliable classification of distress-related facial expressions.

$$F_{i,j}^{(k)} = \sum\nolimits_{m=1}^{M} W_m^{(K)} * X_{i,j}^{(k)} + b^{(k)} \tag{5}$$

$$Y_{i,j} = \sum_{k=1}^{K} W_k * F_{i,j}^{(k)} + b \tag{6}$$

3.5 Pose Estimation-Feature Extraction

MediaPipe Pose is utilized for pose estimation, extracting 33 skeletal keypoints-including the shoulders, elbows, hips, knees, and ankles to capture spatial information essential for posture analysis in distress detection. Once extracted, the system computes joint angles between selected landmarks using the law of cosines, which estimates angles based on the distances between three key points, as expressed in Eq. (2). These calculated angles are then compared against predefined thresholds to classify postures: a hip–knee–ankle angle greater than 170° indicates standing, while an angle below 90° suggests crouching or squatting; a shoulder–hip inclination greater than 45° corresponds to lying down; and an elbow angle less than 90° is interpreted as hands covering the face—a gesture potentially signaling distress.

3.6 Distress Detection

Unlike traditional machine learning classifiers, the proposed system employs a rule-based approach to determine distress by combining the results of emotion recognition and pose estimation. The classification logic is implemented through predefined conditions that check for distress-related behaviors: if both pose and emotion detection fail or if no face and no significant pose are detected, the system classifies the case as No Distress; if emotions such as fear, sadness, or anger are detected but the pose is unknown, it is marked as Potential Distress; if these emotions coincide with a distress-related posture like crouching or covering the face, the system classifies it as Distress: Needs Help; when the emotion is normal but the pose suggests distress, it is flagged as Potential Distress; and if both emotion and posture indicate normal conditions, it is labeled as No Distress. This rule-based framework improves interpretability and reduces the likelihood of misclassification caused by data bias. Upon detecting distress, the system instantly issues an alert through Telegram to pre-registered emergency contacts, ensuring prompt intervention. By incorporating this real-time response mechanism, the system enhances safety and minimizes reaction delays. This ensures that the system remains adaptable and aligned with real-world emergency requirements, thereby offering a practical and effective solution for enhancing women's safety. Table 1 presents the rule-based logic used to determine distress levels by combining emotion recognition and pose estimation results.

Table 1. Rule-based distress classification logic

S.No.	Input	Emotion Detected	Pose Detected	Distress Classification
1.		Fear	Hands Covering Face	Distress: Needs Help
2.		Sad	Unknown Pose	Potential Distress
3.		Normal	Raising hands	Potential Distress
4.		Normal	Sitting	No distress
5.		No Face detected	Hands Covering Face	Potential Distress

4 Experimental Analysis and Results

The proposed distress detection system processes and analyzes facial expressions, body posture, and gender to determine distress levels based on predefined classification logic, with each step of the workflow producing system-generated outputs. The process begins with image preprocessing, where input images are resized to 224 × 224 pixels, normalized to [0,1], and passed through MTCNN for face detection, ensuring only the relevant face region is analyzed. For emotion recognition, a fine-tuned MobileNetV2 trained on FER-2013 classifies faces into fear, sad, or angry, while posture analysis is performed using MediaPipe Pose, which extracts 33 skeletal landmarks to infer postures such as standing, crouching, lying down, or covering the face. Data augmentation techniques like random rotations, brightness adjustments, and horizontal flips are applied to improve robustness. Gender classification is performed using DeepFace, adding contextual information to the distress analysis. The system overlays bounding boxes on detected faces and visualizes skeletal keypoints on the body to enhance interpretability and allow users to verify predictions which is shown in Fig. 5.

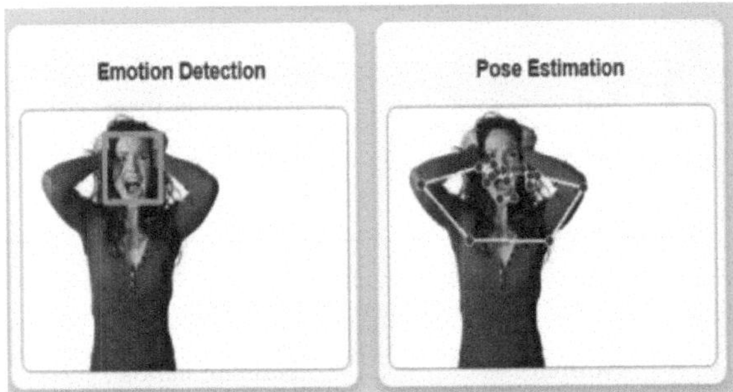

Fig. 5. Emotion detection and pose estimation

In the final stage, a rule-based approach integrates emotion recognition with posture estimation to categorize each scenario into one of three outcomes: No distress, Potential distress, or Distress: Needs help. When both analyses fail, such as in cases of occlusion or detection errors, the system defaults to No Distress. Distress-related emotions (fear, sadness, anger) without distress postures are labeled Potential Distress, while their combination with postures like crouching or covering the face indicates Distress: Needs Help. If only posture suggests distress but emotion is neutral, the case is still flagged as Potential Distress. The system then generates an output showing the analyzed image with detected emotion, posture, gender, and distress classification; in confirmed distress cases, an automated Telegram alert is sent to emergency contacts for timely intervention which is displayed in the Fig. 6.

The experimental evaluation demonstrates the efficiency of the proposed distress detection framework, which fuses facial emotion recognition with pose estimation for practical, real-world deployment. Through the integration of these two modalities, the system can effectively assess distress levels and initiate alerts when required. This multimodal strategy improves the robustness and dependability of distress detection, making it highly applicable to safety and security scenarios.

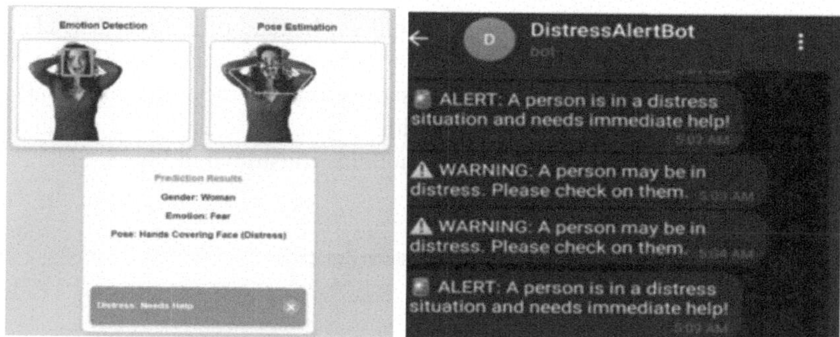

Fig. 6. Distress detection and the alert sent.

5 Performance Analysis

To assess the effectiveness of the proposed system, a comprehensive performance analysis is conducted across its three primary modules: gender classification, emotion recognition, and pose estimation. Each component is evaluated using suitable performance metrics to measure accuracy, robustness, and reliability. The framework integrates DeepFace for gender classification, MobileNetV2 for facial emotion recognition, and MediaPipe for human posture estimation, thereby creating a unified and efficient distress detection system. Unlike conventional single-modality approaches, this multi-modal strategy combines both facial expressions and body posture cues, which enhances adaptability to varied real-world scenarios and improves resilience against input fluctuations.

5.1 Emotion Recognition Model

The emotion recognition module within the multi-modal distress detection framework is based on MobileNetV2, a lightweight deep learning architecture initially pre-trained on the ImageNet dataset and later fine-tuned using the FER-2013 dataset for facial emotion analysis. While FER-2013 originally included seven basic emotions, it was adapted in this study to classify distress-related emotions (sadness, anger, and fear) against non-distress states, where happiness, surprise, and neutral were grouped as "normal". The fine-tuned MobileNetV2 model obtained an accuracy of 71.24%, with a precision of 70.56%, recall of 70.28%, and an F1-score of 70.04%, indicating moderate yet consistent performance. The relatively lower accuracy can be attributed to class imbalance in the FER-2013 dataset, where non-distress samples significantly outnumber distress-related ones. Despite this limitation, the results validate the model's capability to effectively differentiate distress expressions, as depicted in Fig. 7.

5.2 Pose Estimation Model

The pose estimation module of the multi-modal distress detection system employs MediaPipe Pose, a deep learning framework that identifies 33 key skeletal landmarks across the human body. Using these landmarks, the model calculates joint angles to classify postures into categories such as standing, sitting, lying down, as well as distress-associated poses like a defensive stance or slumped shoulders. Instead of a direct ML classifier, predefined threshold-based rules are applied for pose recognition. The model achieved 82.37% landmark accuracy, 74.29% pose classification accuracy, and an overall accuracy of 78.33%, showing its effectiveness in identifying distress-related postures. The performance metrics of the Pose Estimation model is illustrated in Fig. 8.

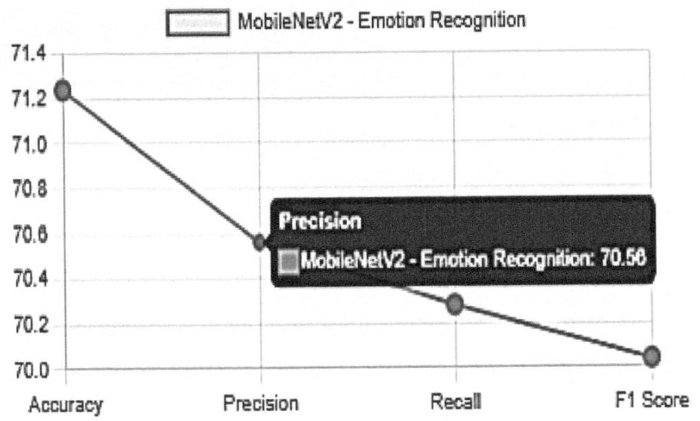

Fig. 7. Emotion Model Performance metrics

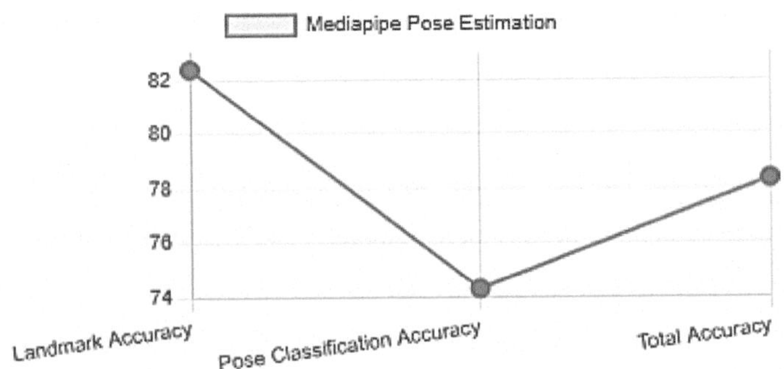

Fig. 8. Pose Estimation Performance metrics

5.3 Gender Recognition Model

The gender recognition module employs DeepFace, a pretrained CNN-based model trained on large-scale datasets like VGG-Face, FaceNet, and OpenFace. It achieved high performance with 97.12% accuracy, 96.29% precision, 95.05% recall, and an F1-score of 95.66%, demonstrating strong reliability across varied conditions. DeepFace's robust feature extraction, face alignment, and pretrained architecture enable efficient, real-time integration into the distress detection system. The performance metrics are shown in the Fig. 9.

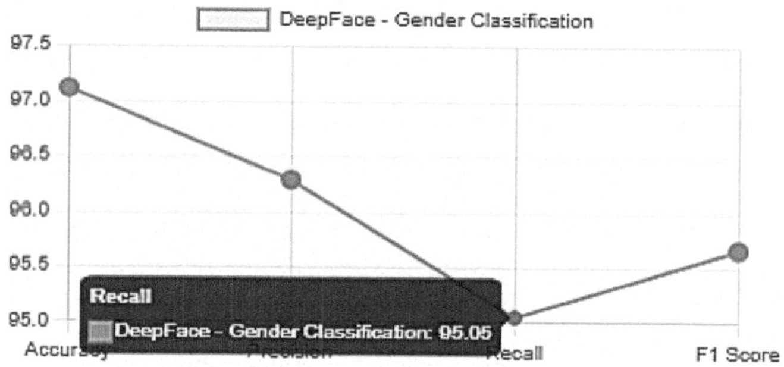

Fig. 9. Gender Model Performance metrics

The multi-modal distress detection system integrates DeepFace for accurate gender classification, MobileNetV2 for reliable emotion recognition, and MediaPipe Pose for posture estimation. Combined, these models leverage both facial and body cues to enhance robustness and adaptability, enabling effective real-time distress detection.

6 Conclusion

This work presents a multi-modal distress detection system that integrates gender classification, facial emotion recognition, and human posture estimation to improve reliability in real-world scenarios. By combining DeepFace, MobileNetV2, and MediaPipe Pose, the framework leverages complementary facial and body cues to accurately assess distress levels. The incorporation of preprocessing, data augmentation, and a rule-based decision mechanism further enhances robustness, minimizing errors caused by data imbalance or environmental variations. Experimental results demonstrate the system's effectiveness in categorizing scenarios into No Distress, Potential Distress, and Distress: Needs Help, while the real-time alert mechanism ensures timely intervention. Overall, the proposed approach provides a practical and efficient solution for safety-critical domains such as surveillance, healthcare, and emergency response.

Discussion

Although the proposed multi-modal distress detection system demonstrates strong performance, future enhancements could further improve its accuracy and robustness. MobileNetV2 may benefit from dataset balancing, advanced fine-tuning, and integration with temporal models like LSTMs to better capture sequential patterns. MediaPipe Pose could be refined to handle occlusions, camera angle variations, and body diversity, potentially through adjusted thresholds or hybrid ML-based classification. DeepFace, already highly accurate and efficient, could be extended to more diverse datasets or combined with additional facial cues for improved distress recognition. These improvements would strengthen the system's adaptability and reliability in complex, real-world scenarios.

References

1. Ali, M, Hussain, A., Sadiq, A.: Human body posture recognition approaches: a review. Aro-the Sci. J. KOYA Univ. **10**. 75–84 (2022). https://doi.org/10.14500/aro.10930

2. Zhang, L., Sun, Y., Wang, M., Pu, Y.: [Retracted] wearable product design for intelligent monitoring of basketball training posture based on image processing. J. Sensors **2021**, 8964433, 15 (2021)
3. Huang, X., Xu, J., Zheng, W., Mao, Q., Dhall, A.: A Survey of Deep Learning for Group-level Emotion Recognition (2024). https://doi.org/10.48550/arXiv.2408.15276
4. Jonathan, J., Lim, A.: Emotion recognition on FER-2013 face images using fine-tuned VGG-16. Adv. Sci. Technol. Eng. Syst. J., **5**, 315–322 (2020). https://doi.org/10.25046/aj050638,
5. Lee, M.-F.R., Chen, Y.-C., Tsai, C.-Y.: Deep learning-based human body posture recognition and tracking for unmanned aerial vehicles. Processes **10**, 2295 (2022)
6. Li, Q, et al.: Real-time facial emotion recognition using lightweight convolution neural network, 012130. IOP Publishing (2021)
7. Liu, G., et al.: Human Posture Recognition Using Kinect Sensors by Advanced Graph Convolutional Netw., 01–07 (2022)
8. Liu, D., Wang Z., , Wang L., , Chen, L.: Multi-modal fusion emotion recognition method of speech expression based on deep learning. Front. Neurorobotics **15** (2021)
9. Chen, Q., Zhang, C., Liu, W., Wang, D.: SHPD: surveillance human pose dataset and performance evaluation for coarse-grained pose estimation. In: 25th IEEE International Conference on Image Processing (ICIP), Athens, Greece, pp. 4088–4092 (2018)
10. Reghunathan, R.K., Ramankutty, V.K., Kallingal, A., Vinod, V.: Facial expression recognition using pre-trained architectures. Eng. Proc. **62**(1), 22 (2024)
11. Urnisha, N., Bithi, S., Rafee, M., Remon, N., Hasan, M., Roy Chowdhury, R.:. A transfer learning approach for facial emotion recognition using a deep learning model. Inter. J. Res. Sci. Innovation XI. 274–284. (2024)
12. Sachin, P., Correa, N., Shenoy, A.H., Ballal, A. C., Mittal, P.: Gender and emotion classification by hierarchical modelling using convolutional neural network. In: 2022 2nd Asian Conference on Innovation in Technology (ASIANCON), Ravet, India, pp. 1–6 (2022)
13. Liaqat, S., Dashtipour, K., Arshad, K., Assaleh, K., Ramzan, N.: A hybrid posture detection framework: integrating machine learning and deep neural networks. IEEE Sensors J. **21**(7), 9515–9522 (2021)
14. Tang, H.-Y., Tan, S.-H., Su, T.-Y., Chiang, C.-J., Chen, H.-H.: Upper body posture recognition using inertial sensors and recurrent neural networks. Appl. Sci. **11**(24), 12101 (2021)
15. Risfendra, Aripriharta, A., Suherman, Ananda, G.F., Putra, D.S., Fahmi, F.: Contactless infant height measurement for enhanced early detection of stunting using computer vision techniques. IEEE Access **1** (2025)
16. Dataset. https://www.kaggle.com/datasets/msambare/fer2013 , Accessed 31 Aug 2025
17. Multi-task Cascaded Convolutional Neural Network (MTCNN), https://medium.com/the-modern-scientist/multi-task-cascaded-convolutional-neural-network-mtcnn-a31d88f501c8
18. Allam, H., Davison, C., Kalota, F., Lazaros, E., Hua, D.: Ai-driven mental health surveillance: identifying suicidal ideation through machine learning techniques. Big Data Cognitive Comput. **9**(1), 16 (2025)
19. Srinivas, S. et al.: Watchguard: real time women safety detection system. In: Goel, S., Sinha, S., Gupta, A. (eds.) AICON 2025. LNICST vol. 672. Springer, Cham (2026). https://doi.org/10.1007/978-3-032-14805-6_7
20. Zhang, T., et al.: Recent advances in video analytics for rail network surveillance for security, trespass and suicide prevention—a survey. Sensors **22**(12), 4324 (2022)
21. Mo, H., Hui, S.C., Liao, X., Li, Y., Zhang, W., Ding. D.: A multimodal data-driven framework for anxiety screening. IEEE Trans. Instrument Measurem. **73**, 1–13 (2024)
22. Asha, P., Balaraman, R.,Prasad, J., Yadav, S., Srinivasan, C.: SVM-based emotion recognition for detecting early signs of psychological distress. In: 2025 5th International Conference on Soft Computing for Security Applications (ICSCSA), pp. 1891–1896. IEEE (2025)

EAMO: An Algorithm for Exploration Among Movable Objects Using a Ground Robot

Nitin Kumar Dhiman[1(✉)], Aniket Kalra[1], and Deepak Khemani[2]

[1] IIT Madras, Chennai, India
nitinkdhiman@gmail.com
[2] Plaksha University, Sahibzada Ajit Singh Nagar, India
deepak.khemani@plaksha.edu.in

Abstract. This paper introduces EAMO (Exploration Among Movable Obstacles), an exploration algorithm designed to reason about the proactive manipulation of objects to access environmental regions previously obstructed by movable obstacles. By incorporating the premise that obstacles can be relocated to reveal hidden areas, the algorithm enables robotic systems to systematically identify and reach regions previously occluded or obstructed by movable objects. The main contributions of this work include: (*i*) a method for representing viewpoints in free regions, a viewpoint corresponding to a movable object, and viewpoint selection; (*ii*) computation of region boundaries (termed gateways) based on the spatial configuration of objects; and (*iii*) representing the exploration output as a sequence of actions. Simulation results indicate that the proposed approach achieves greater coverage compared to the TARE algorithm that does not engage in environmental interaction.

Keywords: Autonomous robotic exploration · coverage · path planning

1 Introduction

Search and rescue, area coverage, and inspection of unknown environments are among the common applications of autonomous robots [2]. Exploration refers to a system's capability to identify a viewpoint location or a path (sequence of viewpoints) leading toward unknown regions to build awareness of the complete environment as rapidly as possible [18]. A viewpoint is a location that the robot can reach to perceive its surroundings or perform an action, granting access to areas for improved spatial awareness of obstructed regions. Most exploration approaches assume that the robot will only perceive, and not interact with, its environment. However, such behavior be restrictive in cluttered environments, where improved results may be achieved by interacting with objects along potential paths. For example, the robot may use its payload arm, if available, or its

© The Author(s), under exclusive license to Springer Nature Switzerland AG 2026
A. Kannan et al. (Eds.): ADCOM 2025, CCIS 2947, pp. 51–65, 2026.
https://doi.org/10.1007/978-3-032-26269-1_4

body to move objects and clear a path for movement. From an exploration perspective, an obstructed area may be reached either by removing objects blocking the way or by searching for an obstacle-free path, with the assumption that such a path exists. This work presents an exploration methodology that evaluates the option of moving objects to achieve better coverage of the environment.

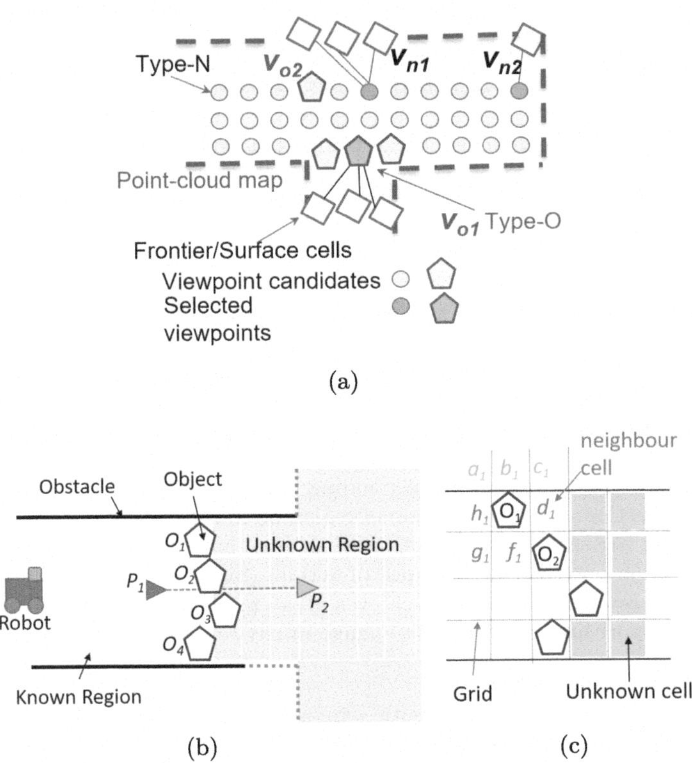

Fig. 1. (a) Type-O and Type-N viewpoint candidates. Type-O corresponds to movable objects inside the planning horizon. The selected viewpoints are shown with an orange colour fill. (b) An example of a gateway: the sequence of objects O_1, O_2, O_3 and O_4 forms a gateway. Point P_1 is identified as the position of gateway entry, which enable manipulation of O_2 and point P_2 defines the destination point into the unknown region after movement of object O_2 (c) immediate neighbourhood cells of the O_1 grid are checked for connectivity with each other. If a path is possible only via a movable object, it is added as part of the gateway.

In the proposed algorithm, viewpoints are categorized into two types (see Fig. 1a): (i) Type-N (navigation type): viewpoints located in free space, where the associated action is perception of the environment; (ii) Type-O (object type): These viewpoints are defined by object locations and are used to enable the robot to enter unknown regions obstructed by objects. The associated action's primary purpose is to move the object, thereby creating a clear passage. The action sequence for Type-O consists of (*a*) navigating to a pose adjacent to the object, (*b*) grasping and relocating the object to a nearby region, if possible, to create a passage, and (*c*) navigating back to the viewpoint pose, followed by traveling to a position into the accessible unknown area. As shown in Fig. 1b, a Type-O viewpoint enables the robot to move from a known region into an unknown region. A sequence of closely placed objects is grouped to define a gateway, representing a transition from a known region into an unknown region. A gateway-based abstraction of the environment allows memoization of the intent behind moving objects, enabling rejection of unfruitful Type-O viewpoint choices in subsequent iterations. After crossing a gateway, the associated objects are not considered as Type-O viewpoints until there is a region accessible only via the movement of those objects.

Figure 1a illustrates two *Type-O* viewpoints, v_{o1} and v_{o2}. The utility of these viewpoints is determined by their marginal spatial gain—the unique volume of unknown space they reveal that is not already perceived by *Type-N* viewpoints. In this instance, v_{o2} is filtered out because it offers no additional spatial gain; the frontier cells it would uncover are already visible from the selected *Type-N* (navigation) viewpoints v_{n1} and v_{n2}. Conversely, v_{o1} is selected as it belongs to a gateway providing exclusive access to a previously occluded region. Furthermore, as depicted in Fig. 1b, multiple objects within a single gateway may share an identical view of hidden frontier or surface cells. However, their overall utility values can vary significantly based on individual grasping and manipulation costs. The formal mathematical framework for these utility computations is detailed in Sect. 4.3. To ensure operational efficiency, our implementation employs a heuristic tie-break that favors the object positioned at the geometric center of the gateway, as relocating a central object typically facilitates the most direct passage for the robot into the newly revealed area.

This article focuses on an exploration algorithm in the presence of known objects in the environment. The following assumptions have been made *w.r.t.* perception and arm manipulation capabilities of the robot: (*i*) the perception of the movable-objects $\mathcal{O}$ is input to the system, which includes attributes of objects *e.g.*, pose, size, affordance *w.r.t.* payload robotic arm *etc.* (*ii*) the robotic arm can grasp any object in $\mathcal{O}$, (*iii*) the payload perception system of the robot can identify a safe place to park/place an object, and (*iv*) joint arm-chassis planning is available, and the robot can execute the plan.

Literature work on embodied visual perception and manipulation [1,19] suggests solutions for building solutions for these assumptions. An volumetric occupancy grid representation of known space in the neighbourhood of the robot can be used for selection of site for relocation of the object. A Behavior-tree [9] based solution combining Nav2 [14] and MoveIt2 [8] can be implemented for robust manipulation and navigation by the robot. For estimating the weight of the unknown objects, robot's arm should have force measurement sensors to estimate the inertial properties and robust execution of plans [15].

The contribution of this work is an exploration planning algorithm for environments with movable objects, and it includes (*i*) categorization of viewpoints and their selection (*ii*) identification of gateways (*iii*) representation of the algorithm output as an action sequence along the local path. Section 2 through Sect. 5 present the related work, problem definition, details of the proposed algorithm, the simulation experiment, and the analysis of results, respectively.

2 Related Work

In [3], the authors present a detailed survey of autonomous exploration algorithms, broadly classifying solutions based on next viewpoint candidate generation, viewpoint selection, and map representation of the environment. These solution approaches passively observe the environment and do not interact with entities to facilitate exploration. The exploration methods using an arm-chassis collaborative approach focus on using the robotic arm as an extended or additional sensor to passively perceive the environment [16,20]. In [10,11], the authors leverage semantic attributes of detected objects to choose custom planning behaviors that aid inspection planning. The proposed approach extends and builds upon the solution provided by TARE [6], a hierarchical exploration method, to handle exploration in the presence of movable objects in the environment. Navigation Among Movable Objects (NAMO) [12,13,17] and embodied navigation [22] utilize an *a priori* known map to plan obstacle-free paths that incorporate manipulation actions to move objects. In [12], the planner computes disconnected regions in the input map and evaluates moving objects to connect these regions, using manipulation when path segments cross movable objects.

In our approach, since an *a priori* map of the environment is unavailable, the algorithm hypothesizes and evaluates gateways based on the current partial map constructed during exploration. A frontier voxel is defined as a free voxel which has unknown voxels in its immediate neighborhood [21]. A surface voxel is defined as a free voxel which has an occupied voxel in its immediate neighbourhood [5]. Onboard sensors, such as the 3D LiDAR, can often perceive through narrow gaps around movable objects, resulting in the formation of these frontier voxels behind the obstruction. These voxels guide the estimation of spatial gain achievable after the relocation of the object. The output of the algorithm is a sequence of actions rather than just a geometric path, capturing the movement actions selected for the involved objects. This representation enables the exploration planner to explicitly communicate action choices to lower-level execution modules.

3 Problem Definition

The proposed approach assumes that the ground robot is equipped with a 3D LiDAR-like navigation sensor S_{nav} with range R_{nav} and 360° field of view, a payload coverage sensor S_{cov} with range R_{cov} and 360° field of view. $R_{cov} \ll R_{nav}$. A voxel is defined as covered if it falls within the line-of-sight and range of the S_{cov} sensor, while it is considered known if it is perceived by the S_{nav} sensor.

A local planning horizon (LPH) is a spatial window with robot at the centre, and the viewpoint selection is performed within this window. Let $\mathcal{O}$ be a set of movable objects in the current local planning horizon window (LPH), which are detected using a perception algorithm employed on a data stream from a detection sensor S_{det} with range of R_{det}. The deployment region considered is a bounded volume $\mathbb{V} \in \mathbb{R}^3$, which consists of many surfaces [6], $\mathbb{S}_{surfaces} = \bigcup_i \mathbb{S}^i_{surface}$ where $i \in 1, 2, 3, ..., n$. The environment, consisting of total volume $\mathbb{V}$, is represented as a grid of $3D$ voxels. $\mathbb{V} = \mathbb{V}_{free} \cup \mathbb{V}_{occ}$. A frontier voxel is defined as a free voxel ($v \in \mathbb{V}_{free}$) that has unknown voxels in its immediate neighborhood.

The algorithm aims to generate a volumetric map of the environment using S_{nav} *i.e.*, all surfaces are mapped, each region is visited, and covered using the sensor S_{cov} and coverage is completed as soon as possible. A voxel is said to be visited or covered if it is within the range R_{cov} and in line-of-sight from any viewpoint in the robot's trajectory.

Let $\mathbb{S}^{covered}_{surface}$ be the surface cells perceived by S_{cov} sensor at a viewpoint till time t. $\overline{\mathbb{S}} = \mathbb{S}_{surfaces} - \mathbb{S}^{covered}_{surface}$ denotes the uncovered cells. A viewpoint $v \in \mathbb{SE}(3)$ defines the pose of the sensor onboard the robot so that awareness and coverage of an unknown environment are obtained. As defined in Sect. 1, a viewpoint may belong to Type-N or Type-O category. The set of active viewpoints L^t_{active} at time t consists of viewpoints which are not visited by the robot and there is unseen area to be observed.

`Problem`: Given $\overline{\mathbb{V}}$, $\overline{\mathbb{S}}$, set of movable objects $\mathcal{O}$, R_{nav}, R_{cov}, R_{det}, current pose of the robot p_{robot}, the set of active viewpoints L^t_{active}, find a sequence of action $P = \{n_1, n_2, ..., n_n\}$ for the robot, when followed, leading to the observation of surfaces in $\overline{\mathbb{S}}$ using S_{cov} and unexplored volume $\overline{\mathbb{V}}$ using S_{nav}. The action $n_i \in P$ either refers to a visit of viewpoint v_i of Type-N or performing a composite action for v_o, a Type-O viewpoint, which involves moving the object $o \in \mathbb{O}$ out of the way and visiting the position of v_o and see ahead of v_o.

4 Proposed Approach

The exploration process can be intuitively summarized as the periodic execution of the following sequence of tasks:

1. *Spatial Representation:* Construct a 3D occupancy and coverage representation based on real-time sensor data.
2. *Viewpoint Selection:* Identify potential viewpoints and evaluate the necessity of moving objects to access obstructed regions using the current partial map.
3. *Path Planning:* Form an efficient path to visit selected viewpoints. For EAMO, this involves identifying the specific action (navigation or manipulation) associated with each viewpoint.
4. *Execution:* Send the action associated with the first node in the path to the robot's execution modules.

The challenges in selecting an exploration approach to solve the above mentioned tasks for unknown and cluttered environments containing movable obstacles include: (i) Limited compute capability: As the spatial map grows from an empty initial state, the computational burden of planning increases significantly. (ii) Dynamic map updates: Because the map is updated online with new sensor data, previously selected viewpoints or paths may quickly become invalid or suboptimal. (iii) Accessibility reasoning: Identifying which parts of the environment are reachable only through the repositioning of obstacles.

The proposed EAMO algorithm addresses these challenges using a hierarchical, receding-horizon planning approach. This hierarchical planning operates at two levels: local and global. Because the robot's map is constantly updating, local planning is executed at a high frequency within a restricted spatial window around the robot, termed the *Local Planning Horizon* (LPH). This ensures reactive and optimal decision-making based on the most recent spatial perception. Conversely, global planning is performed on a lower-resolution map covering the entire environment to guide the robot toward unexplored regions outside the LPH. Global planning partitions the environment into subspaces categorized by three states: *unexplored* (no known spatial data), *exploring* (contains active viewpoints from $\mathcal{L}^t_{active}$), and *explored* (coverage complete). Details of the global planning are presented in Sect. 4.4.

The core logic of EAMO is divided into three interconnected algorithms that function as a Manager, a Scout, and a Pilot:

Algorithm 1 (The Manager): This algorithm provides the high-level control loop for the task cycle. It maintains the coverage map ($M_{coverage}$) and the distance roadmap graph ($\mathcal{R}$), which captures reachable connectivity between voxels to aid efficient path computation. While navigation tasks allow for rapid re-planning, if the Manager commits to an object-movement action, the next planning cycle initiates only after that physical action is fully completed.

Algorithm 2 (The Scout): This algorithm performs viewpoint selection within the LPH. It first prioritizes easy-to-reach navigation spots (*Type-N*). If regions remain hidden, it identifies gateways—clusters of objects that, if relocated, reveal frontier voxels "glimpsed" through narrow gaps. These are transformed into *Type-O* viewpoints, enabling the robot to reason about accessibility through manipulation.

Algorithm 3 (The Pilot): Once viewpoints are selected, the Pilot organizes them into an efficient action sequence. This sequencing is formulated as a Traveling Salesman Problem (TSP) to find the shortest total path. Since the TSP is NP-hard, EAMO computes an approximate solution using the Lin-Kernighan-Helsgaun (LKH) heuristic to maintain real-time performance. The Pilot then links each viewpoint to its specific task: *Type-N* triggers a standard Navigation-Action, while *Type-O* triggers a composite Object-Action involving grasping and relocation.

By linking the "Scout" (selection) to the "Pilot" (sequencing) through the "Manager" (coordination), EAMO ensures every robotic movement is driven towards spatial gain. The pseudocode and technical details for these algorithms are presented in the following sections.

4.1 High Level Control

Algorithm 1 is the main high-level control loop of the task cycle. It run the receding horizon cycle from Line 5 to Line 31.

Algorithm 1. Exploration Among Movable Objects (EAMO) Algorithm

Require: p_{robot}, LPH, SLAM point-cloud map M_{pc}, 3D coverage map $M_{coverage}$
Ensure: Next action for robot, updated $M_{coverage}$

1: // — **Initialization and local setup** —
2: Initialize sets: $L \leftarrow \{\}$; $last\_action \leftarrow NULL$
3: Compute set of movable objects O within the Local Planning Horizon (LPH)
4: Initialize 3D coverage memory $M_{coverage}$ (set all cells to *unvisited*)
5: **loop** // — **Receding Horizon Exploration Iteration** —
6: // **1. Perception: Update spatial and connectivity models**
7: $UpdateCoverageMap(M_{coverage})$; $UpdateDistanceRoadmap(R)$
8: // **2. Selection (The Scout): Identify exploration goals**
9: $L \leftarrow$ SelectViewpoints(p_{robot}, LPH, M_{pc}, $M_{coverage}$, O) ▷ via Algo 2
10: // **3. Sequencing (The Pilot): Compute optimal local action path via Algo 3**
11: $action\_sequence \leftarrow ComputeLocalActionSequence(p_{robot}, M_{pc}, \mathcal{V}_{N+O}, L, R)$
12: // **4. Decision Logic: Prioritize local coverage over global coverage**
13: **if** $action\_sequence \neq \{\}$ **then**
14: $action \leftarrow action\_sequence[0]$
15: **else** // **Fallback to global planning if local area is cleared**
16: **if** $A_{global} \neq \{\}$ **then**
17: $action \leftarrow A_{global}[0]$
18: **else**
19: **return** Exploration completed
20: **end if**
21: **end if**
22: // **5. Global Strategy: Maintain connectivity with viewpoint outside LPH**
23: $A_{global} \leftarrow ComputeGlobalActionSequence()$ ▷ Details in Sec. 4.4
24: // **6. Action Dispatch: Execute specialized routines based on viewpoint intent**
25: **if** $ViewpointType(action) =$ Type-N **then**
26: $SendNavigationPathToRobot(action)$ ▷ Navigate and observe free space
27: **else** // **Interaction commitment for movable obstacles**
28: $SendCompositeObjectActionToRobot(action)$ ▷ Grasp, relocate, and explore
29: **Wait** until manipulation is complete before starting next cycle
30: **end if**
31: **end loop**

Algorithm 1 updates coverage map $M_{coverage}$ (Line 7) and road-map graph R (Line 7). A road-map is a graphical abstraction of the space, capturing the reachable connectivity between voxels. Since, the viewpoint selection process involves computation of shortest paths, a road-map aids in efficient computation of the paths.

A minimum set of viewpoints is selected using Algorithm 2 (Line 9), which is explained in Sect. 4.2. The viewpoints are arranged in an action sequence, *action_sequence*, using the Algorithm 3 (Line 11). The first element of the sequence is selected for execution by the robot (Line 14).

The *action_sequence* is empty if LPH does not have any more viewpoints, that is, the region within LPH is fully covered. In such cases, the algorithm redirects the robot toward the nearest subspace containing an active viewpoint. The algorithm selects the first action element of the global plan, A_{global} as the next visit goal (Line 17). If A_{global} is empty—indicating that no unexplored subspaces remain in the environment—the algorithm terminates successfully (Line 19)

The selected action is executed by calling an appropriate module based on the type of the action *i.e.*. navigation and perception action for Type-N viewpoint (Line 26) or a composite action for Type-O viewpoints (Line 28). The global path is updated in each iteration based on the current active viewpoints (Line 23).

Algorithm 1 commits to the chosen *action* and waits until execution of the same. For example, the algorithm waits for execution of the composite-action sequence associated with Type-O viewpoint, before running the next iteration of the loop in Algorithm 1. Once the robot has crossed into the unknown region, the gateway corresponding to the Type-O viewpoint is no longer valid, therefore, this Type-O viewpoint is discarded. The algorithm will discover new viewpoints consequent to crossing the gateway in the next iteration of the loop.

As EAMO implements a receding horizon hierarchical planning approach, the local path planning is performed at a finer resolution grid in a defined local planning horizon window LPH. Since the size of LPH is bounded, local planning can be computed at higher rate to incorporate spatial updates enabled by newer spatial awareness. Algorithm 3 defines the local path action planning process.

4.2 Selection of Viewpoints

Algorithm 2 details the viewpoint selection process. A minimal set of viewpoints are selected to cover $\mathcal{U}_{\mathcal{LPH}}$, the unknown volume within the LPH. The selected viewpoints may consists of Type-N and Type-O viewpoints. The viewpoints are selected based on the increasing order of the utility value (Eq. 1), with preference for Type-N viewpoints. Type-N viewpoints are selected first (Line 4–9) and Type-O viewpoints are selected to cover the remaining volume $\mathcal{U}_{LPH}^{copy}$ (Line 10–12). The computation of utility of a viewpoint is described in Sect. 4.3.

The viewpoint selection process involves generating the viewpoint candidates of Type-N and Type-O. As shown in Fig. 1a, Type-N candidates are uniformly distributed within traversable space $\mathcal{C}_{trav}^{LPH}$ inside the local planning window LPH. The $\mathcal{C}_{trav}^{LPH}$ corresponds to the space, where the robot can physically reach. The coverage of each Type-N viewpoint is computed, and viewpoints

Algorithm 2. SelectViewpoints: Viewpoint selection algorithm

Require: Robot pose p_{robot}, Local planning Horizon LPH, SLAM point-cloud map M_{pc}, $M_{coverage}$, $\mathcal{O}$ (movable objects in LPH)
Ensure: Set of viewpoints $\mathcal{V}_{N+O}$
1: Compute uncovered volume $\mathcal{U}_{LPH}$ inside the LPH
2: Initialize priority queues $PQ_N \leftarrow \{\}, PQ_O \leftarrow \{\}$; Create map copy $\mathcal{U}_{LPH}^{copy} = \mathcal{U}_{LPH}$
3: // **1. Type-N viewpoint selection**
4: Generate Type-N candidates $\mathcal{V}_N$ within traversable space C_{trav}^{LPH}
5: **for** $\forall v \in \mathcal{V}_N$ **do**
6: Estimate coverage S_v in $\mathcal{U}_{LPH}$
7: Push v to PQ_N with utility (Eq. 1) as priority
8: **end for**
9: Select a minimum subset $\mathcal{V}_{min\_N} \subset PQ_N$ providing maximum utility
10: Update $\mathcal{U}_{LPH}^{copy}$ by removing volume covered by $\mathcal{V}_{min\_N}$
11: // **2. Type-O viewpoint selection**
12: Compute Gateways G by clustering adjacent movable objects $\mathcal{O}$
13: **for** $\forall g \in G$ **do**
14: Generate Type-O viewpoint candidates v_o for the gateway
15: $\mathcal{V}_O \leftarrow \mathcal{V}_O \cup \{v_o\}$
16: **end for**
17: **for** $\forall v \in \mathcal{V}_O$ **do**
18: Estimate potential coverage in the remaining uncovered cloud $\mathcal{U}_{LPH}^{copy}$
19: Push to PQ_O with priority computed using Eq. 1
20: **end for**
21: Select a minimum subset $V_{min\_o} \subset PQ_O$ providing maximum coverage gain
22: Update active global set: $L_{active} \leftarrow L_{active} \cup \mathcal{V}_{min\_o} \cup \mathcal{V}_{min\_N}$
23: **return** $\mathcal{V}_{N+O} = \mathcal{V}_{min\_N} \cup \mathcal{V}_{min\_o}$

are arranged in a priority queue PQ_N based on their utility value (Line 7). A minimal set of Type-N viewpoints, $\mathcal{V}_{min\_N}$, is selected, to maximally cover the unknown volume in LPH (Line 9). The best viewpoints from PQ_N are selected through an iterative process. Because of coverage-overlaps between viewpoints, utility of each viewpoint in PQ_N is recomputed consequent to selection of top element of PQ_N.

Line 10 updates $\mathcal{U}_{LPH}^{copy}$ to mark the voxels covered by PQ_{min_N} viewpoints. Type-O viewpoints are selected to cover the remaining uncovered volume $\mathcal{U}_{LPH}^{copy}$. This process avoids the double counting of unknown voxels towards coverage, which can happen for voxels visible from both a Type-N viewpoint and a Type-O viewpoint.

Type-O viewpoint candidates are generated corresponding to voxels that are obstructed by movable objects, therefore not covered using Type-N viewpoints. The movable objects are abstracted into gateways (Line 12). Please see Fig. 1b for a pictorial depiction of gateway generation. The voxels occupied by the movable objects are clustered into gateways. The neighborhood of the voxels corresponding to movable objects (starting with object O_1's neighborhood $a_1, b_1, ..., h_1$) is evaluated for the existence of a traversable path between them through free voxels. The neighborhood voxels corresponding to the movable objects are checked for traversability by the robot using the robot's footprint. As shown in Fig. 1c, a path around O_1 is possible only via voxel corresponding to other movable object (O_2). The movable object O_2 is clubbed with O_1 as part of this gateway. The extended neighborhood consisting of O_1 and O_2 is further evaluated for traversability. This iterative process could result in two possibilities: (*i*) a traversable path exists in immediate neighborhood of the movable objects. The current set of movable objects is rejected as a gateway candidate. These objects are not considered for Type-O viewpoint generation. (*ii*) The only possible path requires movement of the objects. In this case, The current set of movable objects are accepted as a gateway.

For each gateway, a Type-O viewpoint is generated. A set of Type-O viewpoint candidates, $\mathcal{V}_O$, one for each gateway is computed (Line 15). A utility value for each viewpoint in $\mathcal{V}_O$ is computed using Eq. 1. A minimal set of Type-O viewpoints, $\mathcal{V}_{min\_O} \in PQ_O$ is selected for maximal coverage. Since coverage of a viewpoint may overlap with other viewpoints, the utility of each viewpoint is recomputed consequent to selection of a viewpoint from PQ_O, resulting in update of PQ_O. This is implicitly mentioned in Line 9 and 21. $\mathcal{V}_{N+O}$ is the union of $\mathcal{V}_{min\_N}$ and $\mathcal{V}_{min\_O}$ (Line 23), representing the resultant minimum set of Type-N and Type-O viewpoints which will cover the LPH. L_{active} is also augmented with the selected viewpoints, which is used in sub-space selection for global planning.

4.3 Utility of a Viewpoint

The utility of a viewpoint v_o is computed using Eq. 1. Based on the footprint of the robot, multiple objects ($\mathbb{O}'$) may be required to be relocated from the corresponding gateway. $\forall o_i \in \mathbb{O}'$, the cost consists of three components: (*i*) C_1^i: travel cost as time from the current robot position p_{robot} to the position of o_i (*ii*) C_2^i: cost (time) of positioning robot near the object to enable grasp, grasping an object o and relocating object at an identified free location, and subsequently (*iii*) C_3: travel time to location P_2. β_1 and β_2 hyperparameters normalise the cost of navigation with the cost of manipulation.

The function $UnknownViewableCells(v_i)$ estimates the spatial gain as the set of unknown voxels that would potentially become visible to the robot from the viewpoint v_i. However, in the case of a Type-O viewpoint (v_o), the region beyond the movable object may be occluded. To address this, the algorithm approximates the spatial gain by identifying frontier voxels—free voxels adjacent to unknown space—that are detectable through narrow gaps around the object. Specifically, for a Type-O viewpoint, $UnknownViewableCells(v_o)$ is calculated as the count of these detected frontier voxels located within the Euclidean distance (navigation sensor range, R_{nav}). For a Type-N viewpoint, $UnknownViewableCells(v_i)$ is calculated as the total count of frontier and surface voxels that fall within the line-of-sight of the viewpoint v_i. Specifically, this includes frontier voxels within the navigation sensor range R_{nav} and surface voxels within the coverage sensor range R_{cov}.

$$Utility(v_o) = \alpha_1 * |UnknownViewableCells(v_o)| + \alpha_2 * e^{-C_{manipulation}} \quad (1)$$

$$C_{manipulation} = (\sum_{i \in \mathbb{O}'} \beta_1 * C_1^i + \beta_2 * C_2^i) + \beta_1 * C_3 \quad (2)$$

Algorithm 3. ComputeLocalActionSequence: Local path action sequence planning algorithm

Require: p_{robot}, M_{pc}, $\mathcal{V}_{N+O}$, Distance Roadmap graph R
Ensure: Action sequence $\mathcal{A}$

```
A ← {}
T = A Travelling Salesman Problem (TSP) path order for viewpoints in V_{N+O}.
    Compute Distance between viewpoints using R.
loop ∀t ∈ T
    if Viewpoint-type(t) == Type-N then
        A.push_back(Navigation-Action(t.position))
    else Viewpoint-type(t) == Type-O
        A.push_back(Object-Composite-Action(t.position))
    end if
end loop
```

4.4 Global Path Planning

Global path planning is based on the approach in TARE [6]. To maintain progress toward overall exploration goals without excessive computational overhead, global planning is performed on a coarser spatial resolution map that excludes the current LPH. The environment is partitioned into subspaces, which can exist in one of three states: (i) Unexplored: Areas with no previous sensor coverage, (ii) Explored: Areas where coverage is complete, and (iii) Exploring: Subspaces that currently contain active viewpoints ($\mathcal{L}^t_{active}$). The global planner organizes these "Exploring" subspaces using a TSP algorithm to ensure the robot maintains a consistent direction of travel when no local frontiers are available. The resultant path guides the direction of travel, if LPH does not have any active frontier. The distance between two viewpoints is computed on the grid using the road-map R.

5 Simulation Results

Simulation experiments were conducted using AEDE [7] on ROS2 middleware with the Gazebo simulation environment. A differential drive robot with a 0.8 m radius footprint and a top speed of 2 m/s was used. The $3D$ environment included movable objects and static obstacles. Object perception was obtained from

Gazebo ground-truth ROS2-topics. R_{nav}, R_{cov}, R_{det} were set to 10 m, 2.5 m, and 7 m, respectively. The grasping/manipulation cost C_2 was set to 10 for all objects. Hyperparameters α_1, α_2, β_1, and β_2 were configured to 1, −1, 1, and 1, respectively. Manipulation and navigation costs were measured as time duration required to complete the actions.

The performance of the proposed EAMO algorithm is evaluated through a comparative study against the TARE [4,5]. To ensure a controlled and fair comparison, identical parameter configurations were maintained for both algorithms. The key experimental parameters include a coverage sensor range ($kSensorRange$) of 2.5 m and viewpoint visibility thresholds—defined by the minimum cell count required to consider a location a valid candidate—set at $kMinAddPointNumSmall = 10$ and $kMinAddPointNumBig = 20$. The Local Planning Horizon (LPH) was defined as a 40×40 cell spatial window with a resolution of 0.6 m per cell. For global planning, a coarser map resolution of 5 m was utilized.

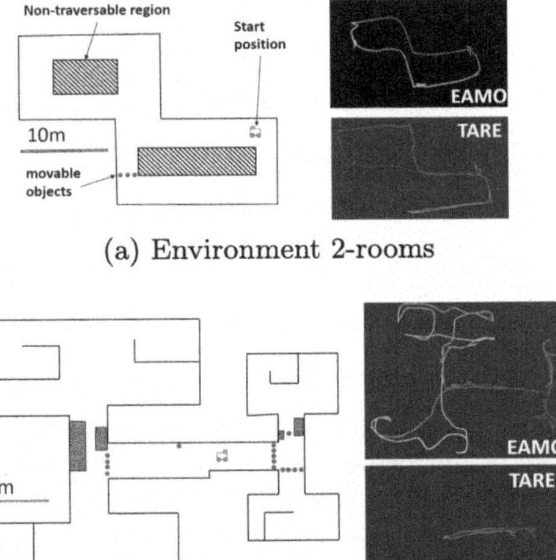

(a) Environment 2-rooms

(b) Environment 4-rooms

Fig. 2. Top-down 2D projection of simulation environment. Trajectory output is presented using the VIBGYOR colour coding. (a) TARE terminates the exploration incompletely as it did not plan for the movement of objects. (b) TARE planned for a longer path to complete the exploration.

The simulation experimentation is performed on two environments. Environment 2-rooms (Fig. 2a) evaluates the capability to consider the object movement in global planning. Environment 4-rooms (Fig. 2b) evaluates the algorithms' capability to update the utility of Type-O viewpoints in local planning, after the robot has crossed into the unknown region.

In the simulation environment '2-rooms', when the TARE algorithm encounters movable objects, they are treated as static obstacles. TARE avoids the movable objects and navigates to the unknown region via the known free region. Figure 2a shows the 120 m long trajectory executed by TARE planner. On the other hand, the proposed EAMO algorithm finds the solution requiring moving the objects to reach the unknown area. EAMO completes the exploration with a 78 mm long trajectory, which is shorter than TARE as shown in Fig. 2a.

In the simulation environment '4-rooms', the starting position of the robot is in a small closed region. The region is bounded on two sides by movable objects, shown as red circles in Fig. 2b. Since TARE does not plan for object movement, it could not come out of the bounded area and terminates the exploration prematurely. EAMO plans for the movement of objects to move outside the bounded region, and completes the exploration. The trajectories of the algorithm are shown in Fig. 2b. TARE does not complete the exploration of the environment.

Table 1 shows that EAMO performs better than TARE in the travel time and trajectory length of the exploration activity for these two environments. The performance gap between TARE and EAMO is rooted in TARE's treatment of movable objects as static obstructions. Consequently, adjusting TARE's parameters will not bridge this gap, as it lacks the inherent reasoning required to evaluate the spatial gain achievable through object relocation.

Table 1. Trajectory length and time spent during exploration runs.

Algorithm	TARE	EAMO
Env-2rooms	120 m, 97 sec	**78** m, **85** sec
Env-4rooms	Incomplete	**480** m, **697** sec

6 Conclusion and Future-Work

This article presents the Exploration Among Movable Objects algorithm, which enables the evaluation of object position as a viewpoint candidate. The viewpoints are categorized as navigation viewpoints and object viewpoints. The association of an action sequence with a viewpoint enables the execution of composite actions at a viewpoint location using Behavior Trees. The viewpoint selection from a cluster of objects is identified using a gateway. The identification of a gateway is also presented. Simulation experiments showed that the proposed algorithm performed better on area coverage and trajectory length criteria.

The current solution approach makes two strong assumptions: (*i*) 360° FoV of the sensors (*ii*) a constant cost of manipulation for each object. Future work will incorporate the visibility constraints of the sensor in viewpoint generation, coverage update, and evaluation of the action choices based on the different operational costs.

References

1. Bai, S., et al.: Towards a unified understanding of robot manipulation: a comprehensive survey. arXiv preprint arXiv:2510.10903 (2025)
2. Bouman, A., et al.: Adaptive coverage path planning for efficient exploration of unknown environments. In: 2022 IEEE/RSJ IROS, pp. 11916–11923 (2022). https://doi.org/10.1109/IROS47612.2022.9982287
3. Brugali, D., Muratore, L., Luca, A.D.: Mobile robots exploration strategies and requirements: a systematic mapping study. Int. J. Rob. Res., 02783649241313471 (2025). https://doi.org/10.1177/02783649241313471
4. Cao, C.: TARE planner. Commit 445007, GitHub repository (2024). https://github.com/caochao39/tare_planner. Accessed Jan 2026
5. Cao, C., Zhu, H., Choset, H., Zhang, J.: TARE: a hierarchical framework for efficiently exploring complex 3D environments. Rob. Sci. Syst. **5**, 2 (2021)
6. Cao, C., Zhu, H., Ren, Z., Choset, H., Zhang, J.: Representation granularity enables time-efficient autonomous exploration in large, complex worlds. Sci. Rob. **8**(80), eadf0970 (2023). https://doi.org/10.1126/scirobotics.adf0970
7. Cao, C., et al.: Autonomous exploration development environment and the planning algorithms. In: 2022 IEEE ICRA, pp. 8921–8928 (2022). https://doi.org/10.1109/ICRA46639.2022.9812330
8. Coleman, D., Şucan, I.A., Chitta, S., Correll, N.: Reducing the barrier to entry of complex robotic software: a moveit! Case study. J. Softw. Eng. Rob. **5**(1), 3–16 (2014)
9. Colledanchise, M., Ogren, P.: Behavior Trees in Robotics and AI: An Introduction, 1st edn. CRC Press (2018). https://doi.org/10.1201/9780429489105
10. Dharmadhikari, M., Alexis, K.: Semantics-aware exploration and inspection path planning. In: 2023 IEEE ICRA, pp. 3360–3367 (2023)
11. Ginting, M.F., Fan, D.D., Kim, S.K., Kochenderfer, M.J., Agha-mohammadi, A.A.: Semantic belief behavior graph: enabling autonomous robot inspection in unknown environments. In: 2024 IEEE/RSJ IROS (2024)
12. He, B., Chen, G., Wang, W., Zhang, J., Fermuller, C., Aloimonos, Y.: Interactive-FAR: interactive, fast and adaptable routing for navigation among movable obstacles in complex unknown environments. In: 2024 IEEE/RSJ IROS, pp. 5402–5409 (2024)
13. Levihn, M., Stilman, M., Christensen, H.: Locally optimal navigation among movable obstacles in unknown environments. In: 2014 IEEE-RAS International Conference Humanoid Robots, pp. 86–91 (2014)
14. Macenski, S., Martín, F., White, R., Clavero, J.G.: The Marathon 2: a navigation system. In: 2020 IEEE/RSJ IROS, pp. 2718–2725 (2020). https://doi.org/10.1109/IROS45743.2020.9341207
15. Mavrakis, N., Stolkin, R.: Estimation and exploitation of objects' inertial parameters in robotic grasping and manipulation: a survey. Robot. Auton. Syst. **124**, 103374 (2020). https://doi.org/10.1016/j.robot.2019.103374
16. Naazare, M., Rosas, F.G., Schulz, D.: Online next-best-view planner for 3D-exploration and inspection with a mobile manipulator robot. IEEE Rob. Autom. Lett. **7**(2), 3779–3786 (2022)
17. Ren, Z., et al.: Search-based path planning in interactive environments among movable obstacles. In: 2025 IEEE International Conference on Robotics and Automation (ICRA), pp. 533–539 (2025). https://doi.org/10.1109/ICRA55743.2025.11128242

18. Stachniss, C.: Robotic Mapping and Exploration, vol. 55. Springer (2009)
19. Tang, Y., et al.: AffordGrasp: in-context affordance reasoning for open-vocabulary task-oriented grasping in clutter. In: 2025 IEEE/RSJ IROS, pp. 9433–9439 (2025). https://doi.org/10.1109/IROS60139.2025.11245995
20. Wang, Y., James, S., Stathopoulou, E.K., Beltrán-González, C., Konishi, Y., Del Bue, A.: Autonomous 3-D reconstruction, mapping, and exploration of indoor environments with a robotic arm. IEEE Rob. Autom. Lett. **4**(4), 3340–3347 (2019)
21. Yamauchi, B.: A frontier-based approach for autonomous exploration. In: Proceedings 1997 IEEE International Symposium on Computational Intelligence in Robotics and Automation CIRA'97. Towards New Computational Principles for Robotics and Automation, pp. 146–151. IEEE (1997). https://doi.org/10.1109/CIRA.1997.613851
22. Zeng, K.H., Farhadi, A., Weihs, L., Mottaghi, R.: Pushing it out of the way: interactive visual navigation. In: CVPR (2021)

Hybrid Method for Automatic Flight Manoeuvre Recognition Using Signal Processing and Classification Learner

Akash Sahu[1], Khadeeja Nusrath[2(✉)], and Dushyant Kaliyari[2]

[1] Bhilai Institute of Technology, Durg, (C.G.), India
[2] CSIR-National Aerospace Laboratories, Bangalore, India
khadeeja@nal.res.in

Abstract. This paper presents a novel hybrid method to automatically detect manoeuvres performed during a flight using signal processing and machine learning techniques. A general framework has been developed that can be applied to all types of flight phases and aircraft. The algorithm is implemented in two steps. Initially, the Continuous Wavelet Transform (CWT) of the signals of interest for different manoeuvres is generated to find the time when a manoeuvre happens during a sortie. CWT looks at the energy levels in the transformed signal and uses smart thresholding to pick out only the parts with high energy to extract real manoeuvres. In the next step, these selected parts are recognised using a pre-trained machine learning model. The pre-trained machine learning model is developed using a supervised learning technique, Random Forest, to recognise different manoeuvres. This approach shows strong promise for use in pilot training in simulators, automatic flight systems, and real-time monitoring. In short, combining continuous wavelet analysis with a classification learner offers an effective way to understand and track how an aircraft behaves during flight using sensor measured data.

Keywords: Flight Manoeuvre Identification · Flight Data Analysis · Wavelet Transform · Random Forest Classifier

1 Introduction

Automated flight manoeuvre recognition (AFMR) offers significant benefits across the aircraft life cycle, from development and testing to operational deployment and training. It enhances safety, reduces manual analysis time, and enables more efficient use of human expertise. During the flight testing phase of an aircraft under development, various manoeuvres are flown as part of a comprehensive flight test matrix by different design and engineering teams. These manoeuvres are essential for analysing the stability, performance, structural behaviour, vibration studies, sensor calibrations, and overall handling quality evaluations. Traditionally, the identification of these manoeuvres is done manually by experts after each sortie, which is not only time-consuming but also requires significant hands-on experience for precise identification. To streamline this process, the

© The Author(s), under exclusive license to Springer Nature Switzerland AG 2026
A. Kannan et al. (Eds.): ADCOM 2025, CCIS 2947, pp. 66–83, 2026.
https://doi.org/10.1007/978-3-032-26269-1_5

development of automated algorithms to recognise test points from flight test data can significantly reduce manual workload. This allows the engineers and analysts to focus more on interpreting results rather than sorting and identifying flight manoeuvre segments. As flight datasets grow in scale and complexity, manual marking becomes impractical, and therefore, automated techniques combining signal-processing heuristics, machine-learning classifiers, and deep-learning sequence models are increasingly adopted to identify manoeuvres such as control doublets, climbs, turns, stalls, and approaches with high temporal precision. Also, in the context of flight vehicle system identification, correctly extracted manoeuvres provide clean and excitation-rich datasets required for accurate estimation of aerodynamic derivatives and control-response models.

Various studies have been conducted on automatic segmentation and flight phase identification. A survey of the application of different deep learning techniques for manoeuvre identification is presented in [1], where the merits and demerits of various deep learning techniques are discussed. A general framework for manoeuvre detection based on clustering and phase space reconstruction is explained in reference [2]. A study, which proposes a machine learning-based framework for detecting and classifying aircraft manoeuvres within air traffic data into three primary manoeuvre types, as detailed in [3]. Machine learning techniques, predominantly outlier detection algorithms, have shown strong performance in recognising manoeuvres in flight data. These methods enable the identification of safety-critical events and abnormal flight patterns with greater accuracy and reliability than traditional techniques [4]. Researchers have demonstrated the effectiveness of the Wavelet Transform in uncovering time-localised patterns associated with various flight manoeuvres, an essential capability for analysing non-stationary signals such as flight data [5, 6], particularly in the context of system identification. This technique provides valuable insights into the time-frequency characteristics of manoeuvres, significantly contributing to comprehensive analyses of aviation incidents and the development of safety improvements. Further investigations have applied the Wavelet Transform to specific manoeuvre events, identifying key features and patterns unique to each manoeuvre type [7]. Wavelets are already being used in aviation to reduce noise in signals or to help predict the aircraft's path [8]. Coming to the classification learner algorithm for aerospace applications, the Random Forest (RF) is applied for the diagnosis of aviation turbulence in [9]. RF approach is used for aircraft trajectory prediction and compared with data-driven models, Long Short-Term Memory (LSTM), and Logistic Regression (LR). The paper concluded with the observation that RF outperforms short-term aircraft trajectory prediction [10].

In previous work by the authors, the LSTM algorithm was employed to recognise manoeuvres, with a wavelet transform applied during the data pre-processing stage to enhance prediction accuracy [11]. While this approach yielded promising results, it required separate training and model development for each manoeuvre, which proved time-consuming. To address this limitation, a hybrid two-step approach was developed to automatically detect manoeuvres in a new flight sortie. In the first step, the continuous wavelet transform (CWT) is applied to detect the presence of manoeuvres during a sortie by analysing the energy content of the measured signals. Thresholds are applied to the wavelet energy of feature signals associated with longitudinal, lateral, and directional motion. This stage only determines whether a manoeuvre has occurred in each motion axis, and it does not classify the specific type of manoeuvre. In other words, the wavelet analysis can indicate that a longitudinal, lateral, or directional manoeuvre took place, but it cannot identify the exact manoeuvre executed. So, in the second stage, an ensemble learning technique based on a random forest classifier is used to categorise the detected manoeuvres into specific classes. The classifier is trained offline using a data bank of manoeuvres collected from previous sorties. The algorithm is now used for automatic manoeuvre extraction and will be extended as a main module for pilot training in simulators. When the module is deployed for in-aircraft training, it must undergo the appropriate certification procedures. However, the governing guidelines for these procedures are still in progress.

The paper is organised into four sections. Section 1 is the introduction. Section 2 details the methodology for manoeuvre extraction using wavelet transform, including the data collection, pre-processing, use of wavelet transforms, energy computation, and thresholding. Details of manoeuvre classification, random forest classifier, model training, and evaluation are also included in this section. Section 3 presents the performance evaluation and limitations of the algorithm, while Section 4 provides a summary and the future scope of work.

2 Methodology

Figure 1 gives the schematic for the automatic flight manoeuvre recognition adopted in the present paper. The block diagram depicts two stages of the algorithm. Initially, the classification learner model is trained using the existing labelled flight test data. Secondly, whenever a new sortie happens, the continuous wavelet transform (CWT) is used to extract all the manoeuvres based on the energy content and thresholding, and the manoeuvres are recognised using the trained model. Each step is explained in detail in this section.

2.1 Data Collection and Data Pre-processing

The development of the AFMR algorithm begins with the collection of flight-measured or simulated data, which can be sourced from previously recorded real-world flights or from aircraft simulators. Modern aeroplanes collect a large volume of sensor data during each flight, including information from GPS, speed sensors, motion sensors, air data sensors, etc. A dedicated flight data bank [12], comprising a range of manoeuvres recorded during actual flights. A part of these data is used to develop the random forest classifier model in the present work. These manoeuvres are initially labelled and stored in a flight management database. Table 1. presents a comprehensive list of manoeuvres used in the present databank.

The database includes a variety of flight conditions such as climb, level acceleration/deceleration, roller coasters (RCOS), wind-up turns (WUT), bank to bank, steady heading side slips (SHSS), and control surface excitation inputs like 3211 sequences and doublets in pitch, roll, and yaw (PDOB, RDOB, YDOB) as given in Table 1. The data consists of both manually flown manoeuvres by pilots and computer-generated pitch, roll, and yaw inputs (PFTP, RFTP, and YFTP) applied through precise control

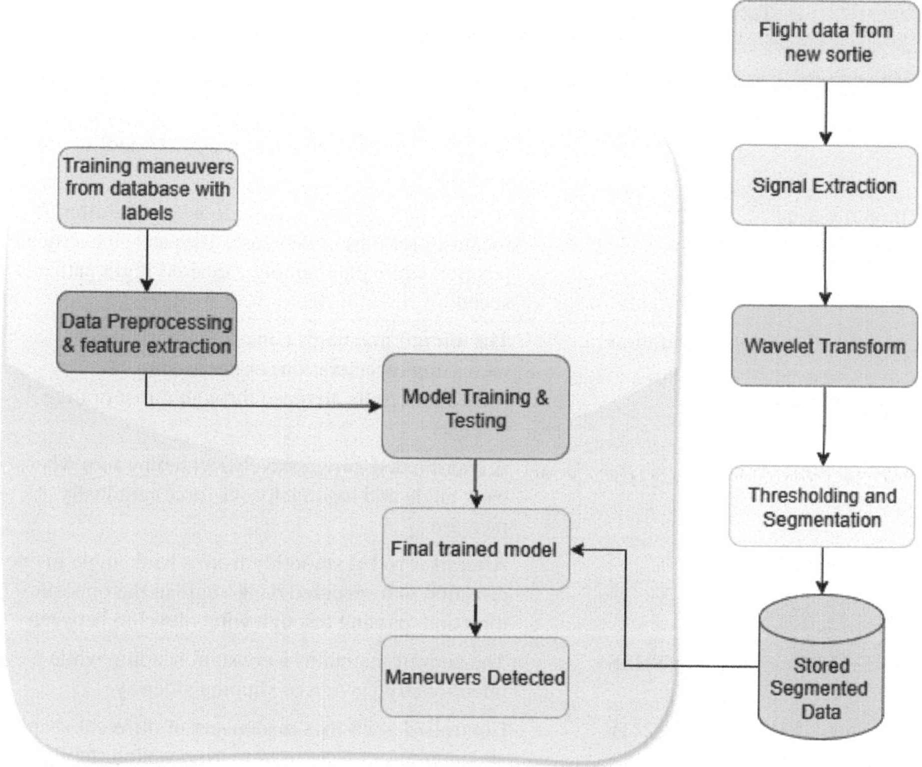

Fig. 1. Flow Diagram for the AFMR

commands. These inputs include sinusoidal signals, pulse inputs, and 3211 or doublet sequences, executed via flight control surfaces such as elevators, ailerons, rudders, and throttles [11, 13]. Uniformly sampled data at 40 Hz were recorded for each maneuver. The typical duration of each maneuver and the number of data files corresponding to each maneuver are summarized in Table 2. Once collected, the data undergoes a pre-processing stage, where noise and missing values are addressed to ensure the dataset is clean, accurate, and complete. Following pre-processing, the relevant signals, along with their nomenclature, are as given below, and features such as mean, standard deviation, min, max, etc., are extracted from these signals for each manoeuvre, preparing the dataset for model training. The maneuvers are labelled with numbers. Table 2 also presents the relevant signals that are used to define the manoeuvres. Typical signals used for manoeuvre identification are

- Angle of attack (Alpha)
- Angle of side slip(Beta)
- True Air Speed(TAS), Calibrated Air Speed (CAS), Mach and Altitude (ALT)
- Elevator (dele), aileron (dela), and rudder (rud) control surface inputs.
- Manual and Computer-generated pilot inputs (PSTK, RSTK, and RPED)
- Euler (Pitch, roll, and yaw) angles and angular rates (phi, theta, psi, p, q, r)

Table 1. Typical Maneuverers in the database with description

Manoeuvres	Description
Roller Coasters (RCOS)	A sequence of alternating pitch-up and pitch-down segments used to assess longitudinal handling qualities
Climb /Descent	A controlled change in altitude where the aircraft increases (climb) or decreases (descent) its vertical position while maintaining a desired flight path and speed
Level Acceleration/Deceleration	The aircraft maintains constant altitude while increasing (acceleration) or decreasing (deceleration) its airspeed through thrust or drag changes
Wind-up turns (WUT)	A coordinated, progressively tightening turn where bank angle and load factor (G-force) gradually increase
Bank to Bank	Aircraft is rolled smoothly from a bank angle in one direction to a specified bank angle in the opposite direction, passing through wings-level in between
Steady Heading Side Slip(SHSS)	The aircraft maintains a constant heading while intentionally yawing or slipping sideways
Pitch input (PDOB/PFTP/3211)	Controlled pitch-axis maneuvers of different shapes and amplitudes were used to assess pitch stability and control
Roll doublet (RDOB /RFTP/Sinusoidal)	Roll-axis maneuvers of different shapes and amplitudes for stability and control testing
Yaw doublet(YDOB/YFTP Sinusoidal)	Yaw inputs of different shapes and amplitudes to excite yaw dynamics
Take-off	Acceleration along the runway and rotation into climb
Landing	Controlled descent, flare, and touchdown followed by rollout

- Translational and rotational accelerations(Ax, Ay, Az, pdot, qdot, rdot)
- Throttle (PLA)
- Weight on wheel (WoW)

Figure 2 shows the concatenated plot of control surface measurements (named as dele,dela,delr in the plot) with the true airspeed measurement for different manoeuvres, such as SHSS, RCOS, RDOB, RPED, 3211, and PDOB. During model training, each of these maneuvers is used as a different file with proper labeling. When similar maneuvers are performed back-to-back during a sortie, they can be processed as a single file. A machine learning model uses this organised and labelled data during training and learn

Table 2. Details of the Maneuverer and the flight data signals used for training

Manoeuvres	Typical Manoeuvre length	No of data set used	Relevant signal that defines the maneuver
Take-off	30-60	189	Alpha, TAS, CAS, Mach, ALT, dele, PSTK, theta, q PLA, Ax, Az, qdot, WoW p, phi ans r are also required for WUT
Landing	30-60	185	
PDOB	5 to 10secs	76	
3211	10 to 12secs	420	
PFTP	5 to 10 secs	503	
RCOS	20to 40 secs	656	
WUT	25 to 50secs	203	
Climb	30 to 60 secs	109	
Level acc/decc	30 to 40 secs	156	
Bank to Bank	20-30	198	TAS, CAS, Mach, ALT, dela, rud, RSTK, RPED, phi, p, r, Ay, pdot, rdot, PLA, Beta
RDOB	10 s	501	
RFTP	10 s	392	
SHSS	15-30	440	
YDOB	10-15	491	
YFTP	10-15	793	
Total manoeuvres		5156	

from these clearly defined manoeuvre segments. Thus, the model becomes better at recognising similar patterns in new, unseen flight data. This careful approach to detecting and labelling manoeuvres helps improve the overall accuracy and reliability of the entire system. Figures 3 and 4 present the time histories of the relevant signals that define the maneuvers SHSS and RCOS. For the roller-coaster maneuver, the principal governing parameters include pitch angle (theta), pitch rate (q), normal acceleration (Az), true airspeed (TAS), angle of attack (Alpha), elevator deflection (dele), throttle setting (PLA), and pilot stick input (PSTK).

Similarly, for the SHSS maneuver, the primary signals are sideslip angle (Beta), true airspeed (TAS), angle of attack (Alpha), rudder deflection (delr), pilot rudder pedal input (RPED), and yaw rate (r)

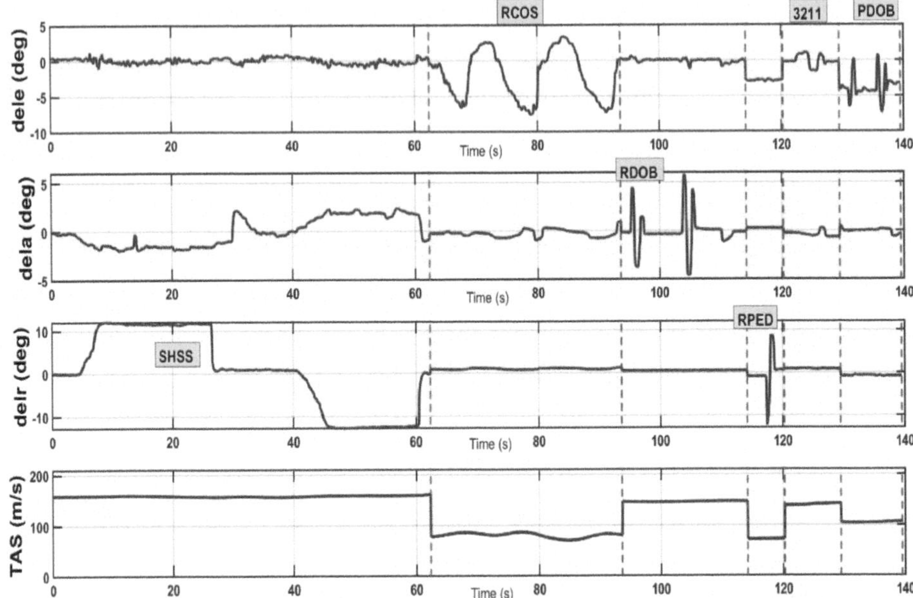

Fig. 2. Typical plot of control surface inputs for different manoeuvres

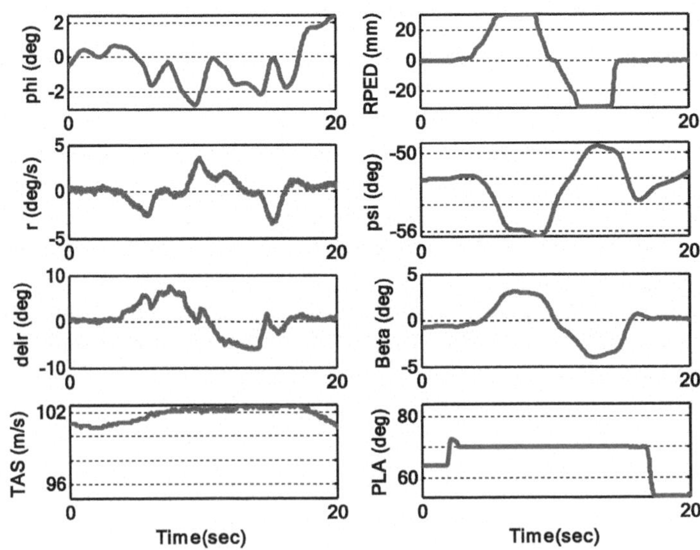

Fig. 3. Time history of flight data signals of SHSS Manoeuvre

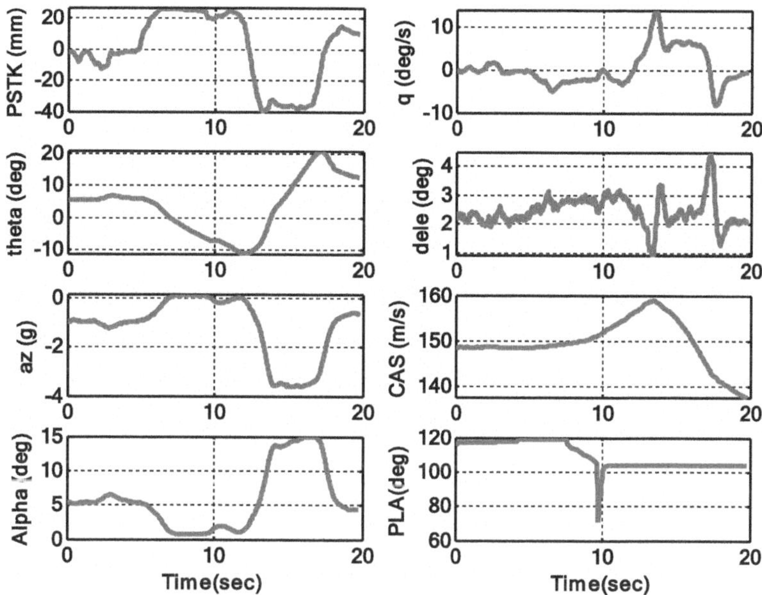

Fig. 4. Time history of flight data signals of RCOS Manoeuvre

2.2 Model Pre-training Using Random Forest Classifier

After collecting the labelled flight manoeuvre segments, the scheme moves on to the feature extraction step. In this stage, various characteristics (features) are calculated from each segment to help the model understand and classify them later. These features include fundamental statistical values such as the mean, standard deviation, skewness, and kurtosis, which describe the shape and behaviour of the signals of interest, providing deeper insight into the strength and complexity of the signal over time. All of these measurements are combined into feature vectors, which are then used as input to the classification model. For the classification part of the system, a machine learning algorithm, a Random Forest classifier, is used [14]. This model is chosen because it is easy to interpret and performs reliably with structured and tabular data. Figure 5 shows the general schematic of the random forest classifier used in the present work. Here, the signals referred to are the flight parameters mentioned in the previous section. A total of 5156 data sets, including all listed manoeuvres, were used for training and testing.

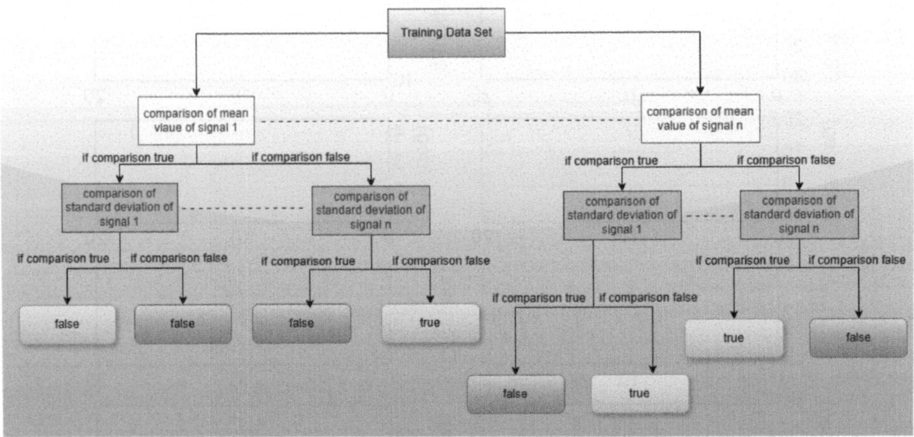

Fig. 5. Example of Random Forest Classifier with mean and standard deviation as features

All the data were combined, and the model training was carried out using 80% of the data. The remaining 20% was used for testing. The Random Forest model works by combining the results of many decision trees, allowing it to learn complex rules from the data and reduce the chances of overfitting, as shown in Fig. 5. This makes it a powerful tool for accurately classifying different types of flight manoeuvres.

2.3 Model Evaluation

To assess the model's performance, it is evaluated using standard classification metrics, including accuracy, precision, recall, and F1 score [15]. These parameters help us comprehend how often the model makes correct predictions, how well it identifies each type of manoeuvre, and how balanced its results are. In addition to these metrics, a confusion matrix shown in Fig. 6 is used to visualise where the model may have made mistakes. In this matrix, results manoeuvres like PFTP, RCOS, RDOB, RFTP, RPED, SHSS and YFTP are considered. For the current model, the accuracy, F1 score, and recall/ precision values are obtained as shown in Table 3.

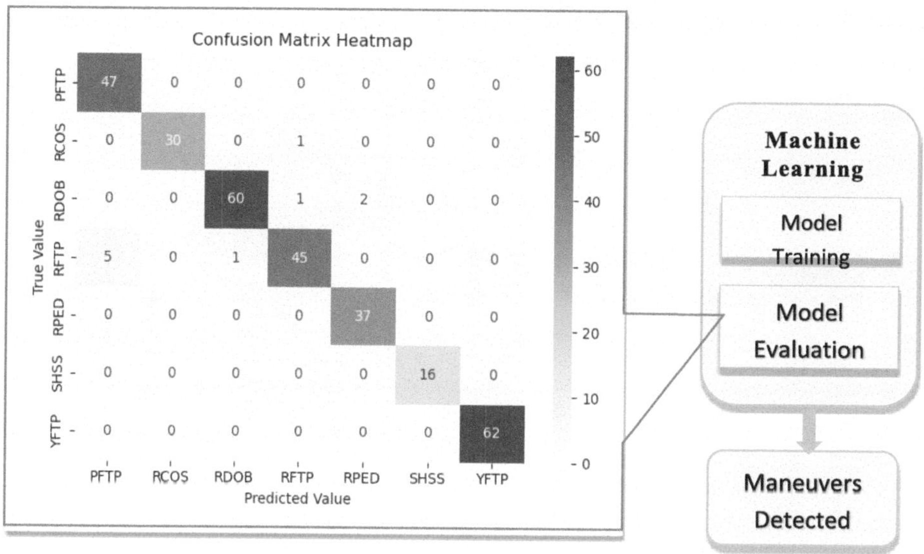

Fig. 6. Confusion Matrix Heatmap

Table 3. Model Evaluation

Manoeuvres	Precision	Recall	F1-Score	Support
PFTP	0.92	0.96	0.94	47
RCOS	0.97	0.91	0.94	32
RDOB	0.94	0.91	0.92	67
RFTP	0.87	0.89	0.88	45
RPED	0.91	0.97	0.94	31
SHSS	1.00	0.94	0.97	16
YFTP	1.00	1.00	1.00	67
Accuracy			**0.94**	**305**
Macro Avg	0.94	0.94	0.94	305
Weighted Avg	0.94	0.94	0.94	305

2.4 Flight Data from New Sorties and Data Signals

The aircraft generates new data every day, which needs to be recognised and extracted for further analysis. Here, the data may contain the manoeuvres from longitudinal, lateral/directional axes. For each of the manoeuvres, the signals that describe them are different, and all these signals need to be extracted to recognise each of them. The data is pre-processed to remove any noise, scale factor, bias, and time shifts. The Continuous Wavelet Transform (CWT) is used to automatically extract the manoeuvre performed during the test flight sortie. Figure 7 shows the schematic for wavelet transform-based segment extraction from the full flight sortie.

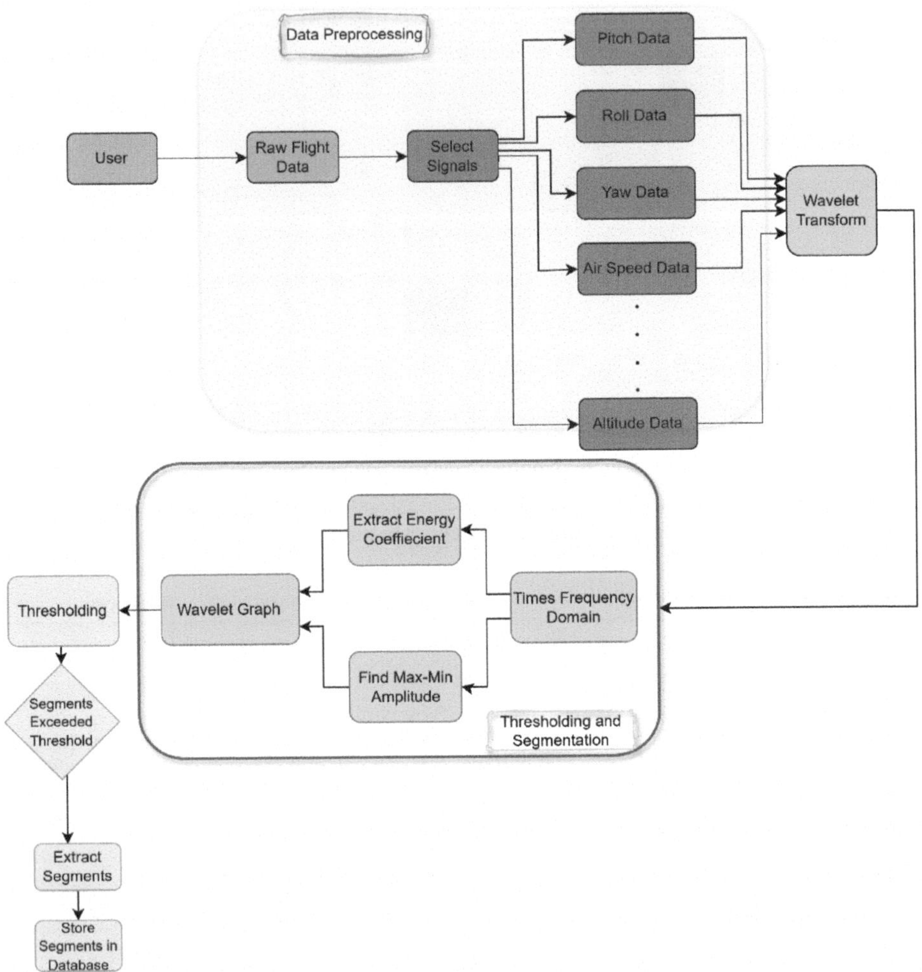

Fig. 7. Segment extraction with wavelet transform

The flight data parameters that define each manoeuvre are recognized, and their wavelet transform is computed. The parameters separately for manoeuvres in pitch, roll, and yaw axes, along with the other relevant parameters, were identified for each manoeuvre. From the continuous wavelet transform coefficients, the total energy coefficients are computed, and the segments are extracted based on thresholding. For example, take off can be recognized using engine power or torque, weight-on-wheels indications, altitude, airspeed, angle of attack variations etc.

2.5 Continuous Wavelet Transformation (CWT)

The Continuous Wavelet Transform (CWT) is a time-frequency analysis tool particularly well-suited for signals whose frequency characteristics vary over time [16]. Unlike fixed-window methods, CWT uses variable-sized windows. The window broadens in time, making it suitable for low-frequency events, and narrows for high-frequency events. This adaptability makes CWT ideal for identifying transient behaviour, sudden frequency changes, and gradually evolving signal characteristics, which are the common traits in flight data. The equation for the continuous wavelet transform is given below.

$$W_x(a,b) = \frac{1}{\sqrt{|a|}} \int_{-\infty}^{\infty} x(t)\psi * \left(\frac{t-b}{a}\right) \mathrm{dt} --- (1) \quad (1)$$

where:

- *x(t)* is the signal being analysed.
- *ψ(t)* is the mother wavelet.
- $\psi * (t)$ is the complex conjugate of the mother wavelet.
- a is the scale parameter (determines the dilation or contraction of the wavelet).
- b is the translation parameter (determines the position of the wavelet in time).
- $\sqrt{|a|}$ is the normalisation factor to ensure energy conservation in the transform.

In the present study, CWT filter banks are used, and the hyperparameters used in a CWT filterbank are scale, signal length, wavelet type, sampling, frequency limits, period limits, and voice per octave. Brief description of each of these parameters are given below.

Scale: Process of stretching or shrinking the signal, a positive value expresses how much the wavelet is scaled in time. A mix of scaling values is used to capture all the frequency ranges in the data.

Signal Length: Length of the signal to be analysed.

Voice Per Octave: How many wavelets are used in an octave. Equivalent frequency of a wavelet is reduced by octaves by scaling by factor of 2. CWT can analyse the intermediate frequency between the octaves called as Voice per octave.

Wavelet Type: Typically used Morse, Morlet and Bump are used.

Frequency Limits: Specified as a two-element vector, representing the highest and lowest peak band frequency.

Period Limits: Period limits of the wavelet filter bank, specified as a two-element duration array. The first element of period limits specifies the largest peak passband frequency. The base-2 logarithm of the ratio of the minimum period, minP, to the maximum period, maxP, must be less than or equal to -1/NV, where NV is the number of voices per octave:

log2 (minP/maxP) ≤ -1/NV.

CWT is particularly effective at isolating non-stationary components in flight data signals. In this work, both Morlet and Morse wavelets were implemented, and each provided satisfactory results, in other words, could successfully extract the occurrence of various maneuvers in the flight data. When applied to flight signals, the CWT generates visual and numerical outputs known as scalograms, which represent the distribution of signal energy over time and frequency. These scalograms help identify periods of dynamic activity, such as during specific flight manoeuvres. After generating scalograms, the system analyses energy distribution patterns (as illustrated in Fig. 8) using a combined energy and entropy-based approach. Entropy, which quantifies signal complexity, helps pinpoint the most informative sections of the signal.

For each flight variable (e.g., pitch, roll), adaptive thresholds are set to distinguish between steady flight and active motion. When the energy of a signal segment exceeds its respective threshold, it is flagged as a potential manoeuvre. Following this detection step, the algorithm categorises the manoeuvre as either lateral (e.g., turns, roll doublets) or longitudinal (e.g., climbs or descents, pitch doublet /3211), and segments the flight data accordingly. Through this method, multiple relevant segments are extracted from new flight data based on the energy profiles across various flight parameters.

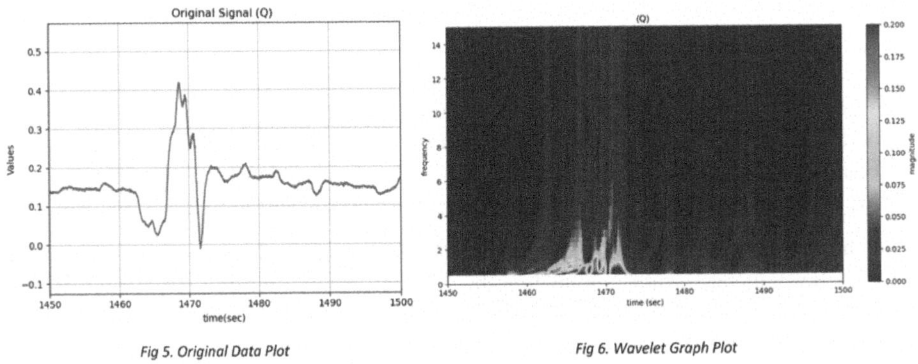

Fig. 8. Normalized pitch rate and its corresponding scalogram

Figure 9(1) shows the manoeuvres in the one flight test sortie marked in yellow, and Fig. 9(2) shows one such manoeuvre extracted. There are many manoeuvres of similar types performed in the flight data as shown in Fig. 9(1). As an example case one maneuver was extracted and shown in Fig. 9(2). All the data above the threshold was considered as a maneuver, and 2 seconds of data before and after the threshold were also considered. The amlplitude was normalized with maximum value and plotted in the Fig. 8 as a typical example. If the data selected is a pitch rate signal in Fig. 9(1), the manoeuvres marked by yellow can be longitudinal manoeuvres such as PFTP, PDOB, RCOS etc. If it represents a yaw rate, it can be YFTP, YDOB, SHSS etc.

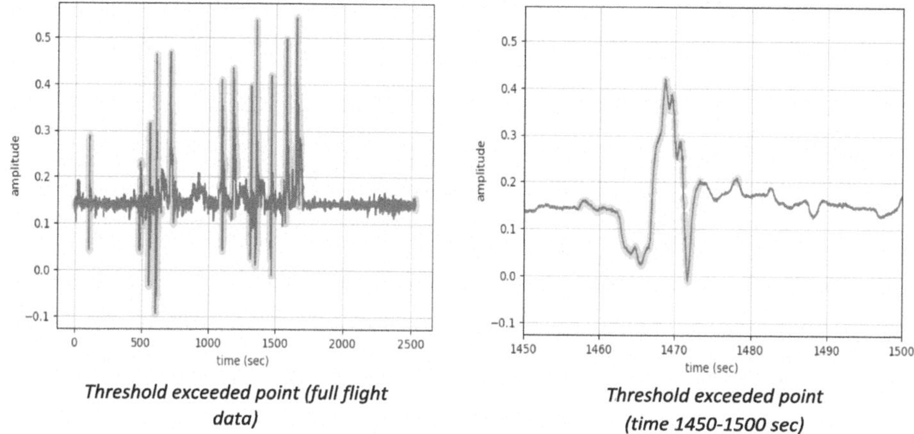

Fig. 9. Threshold exceeded point plot for full flight and extracted maneuverer

This step ensures that each data segment reflects a meaningful flight action rather than just a random slice of time. Once these segments are extracted, they are saved in a structured database along with other data segments. Scale values ranging from 1 to 3 were selected according to the frequency content of the chosen signal. Both Morse and Morlet wavelets were employed in this study. The hyperparameters used for the wavelet analysis included a VoicesPerOctave value of 10, a sampling frequency of 1 Hz, and frequency limits set to [0, 500] Hz. Based on the selected frequency limits, the corresponding period limits were computed using the PyWavelets Python library [17]. Scalograms were generated using the Scaleogram library, which is built on top of PyWavelets [18].

2.6 Manoeuvre Recognition (Model Prediction)

The trained Random Forest model used a combination of features such as the mean, standard deviation, skewness, kurtosis, and other statistical values, which help to pick up on subtle changes in flight behaviour that signal specific manoeuvres like turns, climbs, descents, or level flight. When new data came in, the model automatically ran it through the same steps used during training: cleaning the data, extracting features, and analysing the patterns. Then it applied the rules it had learned to decide which manoeuvre was most likely happening in each segment. The predicted results closely matched the actual manoeuvre labels, showing that the model was very accurate in capturing the complex flight dynamics.

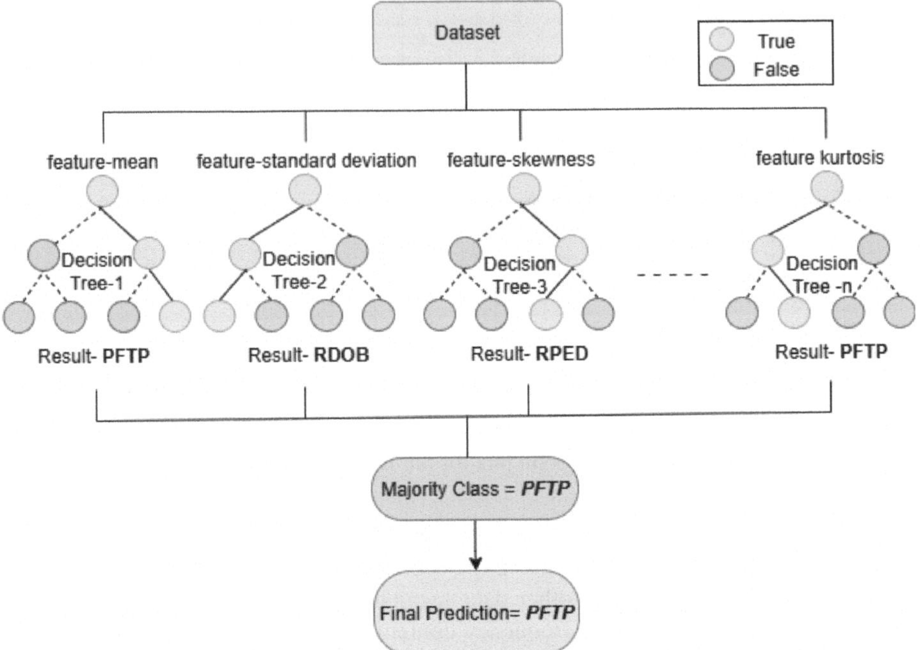

Fig. 10. Random Forest Classifier prediction

These results in Fig. 10 highlight how useful the system could be in real-world settings, whether it's reviewing flight recordings, helping evaluate pilot performance, or even being used in real-time systems onboard aircraft to support decisions during flight. Figure 8 illustrates the identification of Pitch input (PFTP) using a trained Random Forest Classifier. Some of the decision trees identify the manoeuvre as Roll input (RDOB) and Rudder Pedal input (RPED), but the majority identify it as PFTP. So the final prediction will be based on the majority class output.

3 Performance Analysis and Limitations

Tests and experiments (mentioned in Figs. 10 and 11, and Table 3) show that the system can accurately detect and classify common flight manoeuvres using sensor data. By using Continuous Wavelet Transform (CWT), the system can clearly identify instants in the signal where energy levels spike, and these instants often mark important manoeuvres (shown in Fig. 7). Adding both lateral (like roll and yaw) and longitudinal (like pitch) input data helps the model better recognise a wide range of maneuver types, improving its overall accuracy and reliability. The Random Forest model performed well in different test situations, consistently delivering high accuracy even when the data included noise or varied slightly from what it had seen during training. The confusion matrix is the tool that shows which manoeuvres the model got right or wrong, indicating that the system made very few errors, meaning it could effectively tell different manoeuvres apart. In addition, visual tools like feature importance graphs and energy heatmaps help explain how the model makes decisions. This makes the system especially useful in areas where people still need to review or validate the results, such as in pilot training reviews or post-flight performance analysis.

As the Continuous Wavelet Transform is computationally intensive, using the algorithm in real-time situations is more difficult, even though it performs very well in offline testing. While CWT gives detailed insights into how signals change over time and frequency, it is more challenging to implement than other faster methods like the Discrete Wavelet Transform or simpler time-domain filters. Another limitation is that the algorithm relies on labelled training data and needs examples of each manoeuvre with correct labels. Creating these labelled datasets can be difficult, especially if the data is inconsistent or incomplete. Also, the predictor's performance can be degraded when it encounters new or unclear manoeuvres that weren't part of the training data.

Even with these challenges, the algorithm still offers a strong and practical approach to recognising flight manoeuvres using a structured and available data set. It also creates a solid foundation for future improvements and more advanced applications such as intelligent pilot simulators and in-flight pilot training devices.

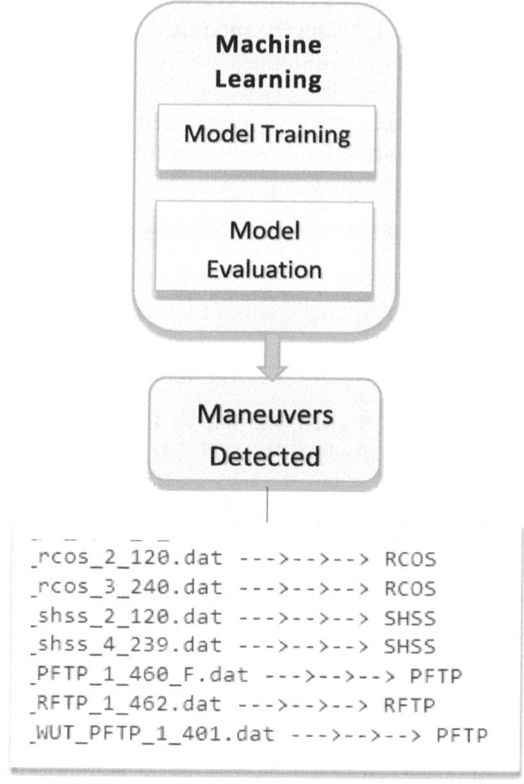

Fig. 11. Model prediction

4 Conclusions

This research presents a general framework for recognising aircraft manoeuvres by combining wavelet-based signal processing with machine learning algorithms. Using the Continuous Wavelet Transform (CWT), the scheme can accurately decompose flight data and detect changes in signals that characterise aircraft movement. These signal changes are then used to classify the manoeuvres using a pre–trained Random Forest classifier, which matches energy-based features to specific types of manoeuvres. More than 5000 manoeuvre segments were used for model training. The results show that this framework works well for analysing flight data after a flight, helping the analyst expedite the process and supporting a smart flight analysis structure for decision-making. The algorithm can be easily configured for any aircraft of a similar category. These models can learn more complex patterns directly from raw sensor data. Also, faster processing methods like lightweight wavelet techniques or parallel computing could be explored to make it work in real-time during actual flights. These upgrades would help make it more flexible, faster, and better suited for use in the next generation of aviation technology. Future work will also aim to explore more powerful classifiers that are suitable for time series data. The integration of the model with aircraft simulation software for AI-enabled pitot training is also to be realised. Using the module for training in a real aircraft will require certification, and the relevant guidelines are still under development.

Acknowledgements. The authors would like to express their sincere gratitude to the Director, CSIR- National Aerospace Laboratories, for providing the resources and support necessary to carry out this research.

Disclosure of Interests: The authors have no competing interests to declare that are relevant to the content of this article.

References

1. Lu, J., Pan, L., Deng, J., Chai, H., Ren, Z., Shi, Y.: Deep learning for flight manoeuvre recognition: a survey. Electr. Res. Archive **31**(1), 75–102 (2023). https://doi.org/10.3934/era.2023005
2. Lu, J., Chai, H., Jia, R.: A general framework for flight maneuvers automatic recognition. Mathematics **10**, 1196 (2022). https://doi.org/10.3390/math10071196
3. Dang, P.H., Tran, P.N., Alam, S., Duong, V.N.: A machine learning-based framework for aircraft maneuver detection and classification. In: Fourteenth USA/Europe Air Traffic Management Research and Development Seminar (ATM2021) (2021)
4. Oehling, J., Barry, D.J.: Using machine learning methods in airline flight data monitoring to generate new operational safety knowledge from existing data. Saf. Sci. **114**, 89–104 (2019). https://doi.org/10.1016/j.ssci.2018.12.018
5. Guo, T., Zhang, T., Lim, E., et al.: A review of wavelet analysis and its applications: challenges and opportunities. IEEE Access **10**, 58869–588903 (2022). https://doi.org/10.1109/ACCESS.2022.3179517
6. Naruoka, M., Hino, T., Tsuchiya, T.: Review of real-time system identification of aircraft dynamics using time-frequency wavelet analysis. In: 27th Congress of the International Council of the Aeronautical Sciences (ICAS 2010) (2010)

7. Wang, Y., Dong, J., Liu, X., Zhang, L.: Identification and standardisation of maneuvers based upon operational flight data. Chin. J. Aeronaut. **28**(1), 133–140 (2015). https://doi.org/10.1016/j.cja.2014.12.026
8. Zhang, Z., Guo, D., Zhou, S., et al.: Flight trajectory prediction enabled by time-frequency wavelet transform. Nat. Commun. **14**, 5258 (2023). https://doi.org/10.1038/s41467-023-40903-9
9. Williams, J.K.: Using random forests to diagnose aviation turbulence. Mach. Learn. **95**, 51–70 (2014). https://doi.org/10.1007/s10994-013-5346-7
10. Hashemi, S.M., Botez, R.M., Ghazi, G.: Robust trajectory prediction using random forest methodology application to UAS-S4 Ehécatl. Aerospace **11**(1), 49 (2024). https://www.mdpi.com/2226-4310/11/1/49
11. Parihar, P., Kumar, U., Kaliyari, D., Tk, K.: Automatic maneuver detection in flight data using wavelet transform and deep learning algorithms. In: SAE Technical Paper 2024–26–0462 (2024). https://doi.org/10.4271/2024-26-0462
12. Nusrath, K.N.T.K.: Automated Flight Data Management System for Aircraft System Identification (AFDMS). FMCD, CSIR- NAL
13. Raol, J.R., Girija, G., Singh, J.: Modelling and Parameter Estimation of Dynamic Systems (1992). https://doi.org/10.1049/PBCE065E
14. Breiman, L.: Random forests. Mach. Learn. **45**, 5–32 (2001). https://doi.org/10.1023/A:1010933404324
15. Manning, C., Raghavan, P. and Schütze, H. (2008) Introduction to Information Retrieval. Cambridge University Press, Cambridge. http://www-nlp.stanford.edu/IR-book, https://doi.org/10.1017/CBO9780511809071
16. Aguiar-Conraria, L., Soares, M.J.: The continuous wavelet transform: moving beyond uni- and bivariate analysis. J. Econ. Surv. **28**(2), 344–375 (2014). https://doi.org/10.1111/joes.12012
17. Lee, G.R., Gommers, R., Wasilewski, F., Wohlfahrt, K., O'Leary, A.: PyWavelets: a python package for wavelet analysis. J. Open Source Softw. **4**(36), 1237 (2019).https://doi.org/10.21105/joss.01237
18. https://pypi.org/project/scaleogram

Energy and Environment

Net Zero Energy Home with Solar Powered Photovoltaic System

Makarand M. Lokhande[1](✉), Prakash S. Kulkarni[1], Pravin D. Sawarkar[2], Ramsha Karampuri[1], Anindita Roy[3], and Raju Lenkalapelly[1]

[1] Department of Electrical Engineering, Visvesvaraya National Institute of Technology, Nagpur, India
mml@eee.vnit.ac.in

[2] Department of Mechanical Engineering, Visvesvaraya National Institute of Technology, Nagpur, India

[3] Mechanical Engineering Department, Symbiosis International (Deemed University), Symbiosis Institute of Technology, Pune, India

Abstract. Inverter interfaced grid connected Rooftop Solar Photovoltaic (PV) Systems (RTS) are increasingly being used for residential, commercial and industrial units. This growth is accelerated by reduced cost of solar PV panels, inverter and availability of government subsidy. Residential and commercial consumers install the system of capacity equal to its sanctioned demand. This is also based on the availability of the area for installation. Even if sufficient space is available at the consumer's premises to install a PV system, the installed capacity cannot exceed the sanctioned demand due to utility regulations. Also, the sanctioned demand is always higher than the actual electrical unit consumption. So annually RTS generates surplus units, which are purchased by the utility. However, the utility purchases it at very low cost. This paper proposes to manage the generated electrical units by RTS for self-consumption for fulfilling all energy needs of the consumer such as water heating, hydrogen cooking and EV charging etc. All these loads can be operated directly from the DC side of the PV system. Case study of the consumer with 3kW sanctioned load is considered with an RTS rating of 5 kWp. It is shown that a 5 kWp system fulfills the total energy requirement of the house. For the proposed system, investigations are carried out at various locations in India.

Keywords: Solar Energy · Photovoltaic System · Thermal Storage · Hydrogen Generation · EV charging

1 Introduction

RTS connected to the grid frequently generates surplus energy due to lower household electricity consumption compared to the installed capacity of RTS. Traditionally, this excess energy is exported to the utility grid. If at any instant, surplus solar generation exceeds the rated capacity of the distribution lines or the transformer, the utility enforces the curtailment of solar energy [1]. This will result in revenue loss for the customer who

© The Author(s), under exclusive license to Springer Nature Switzerland AG 2026
A. Kannan et al. (Eds.): ADCOM 2025, CCIS 2947, pp. 87–98, 2026.
https://doi.org/10.1007/978-3-032-26269-1_6

owns the RTS. Also, the monetary compensation for exported energy to the grid from RTS is significantly lower than the retail electricity cost. Therefore, the owner of the RTS remains dissatisfied because of curtailment of solar energy and with the lower value of compensation.

The best way to address this issue of surplus generation is to increase RTS self-consumption for various household activities. Apart from conventional loads like, fan, lights, refrigeration, air conditioning, one can think of increasing self-consumption by connecting loads like battery storage, electric cookers, hydrogen generation by splitting of water using electrolyzer, and the charging of electric vehicles, electric water heating thermal storage etc. For some countries, the amount of subsidy generally provided by the government is going to be reduced [2]. So increased self-consumption could raise the profit of PV systems and lower the stress on the electricity distribution grid.

Various studies have been reported in the literature under the title of net zero energy (NZE) home, or net zero energy building (NZEB) where most of the cases are related to electrical energy only. The reference [3] proposes a NZE house with a Hybrid PV Energy Storage System that combines a battery with a variable-power Electric Water Heater (EWH) connected to the DC side of the PV system. By coordinating EWH operation with solar generation, the system shifts hot-water heating to midday. This reduces battery charging requirements, and lowers the needed battery size by up to 30%. In [4], for achieving a net-zero home through efficient battery sizing, grid interaction to balance energy generation and consumption an optimal sizing algorithm is implemented and a novel constraint is introduced to enforce net zero transfer. The system displayed a lot of switching between the grid and the BESS to satisfy the load. Net energy home with solar and wind energy systems along with battery and hydrogen energy storage systems is addressed in [5]. When generation is more than demand, energy is stored in Battery Energy Storage System (BESS) and this energy is utilized when needed. The extra amount of energy is used for electrolysis of water for generating hydrogen for internal combustion engine vehicles. The proposed problem is mathematically expressed as mixed integer nonlinear programming and solved using particle swarm optimization technique. Objective function of the problem is to minimize the number of charging-discharging cycles resulting in increasing the life-cycle of BESS. Roy et al. [6, 7] proposed a methodology for sizing and optimization of hybrid wind-solar-photovoltaic system for meeting the electricity demand of isolated rural habitats unconnected to the grid. The method termed as design space also took into account the uncertainty of renewable resource [7]. The photovoltaic hydrogen production potential for industrial zones in Vietnam is analysed in [8]. The Homer software was used to simulate and calculate power output. The results showed that hydrogen can be generated from about 17386 kg/year to 17422 kg/year and can be supplied for operation of fuel cells. Study in [9] examined the application of a solar powered PEM electrolyser system to produce hydrogen at a large scale as a domestic cooking fuel in developing countries. Paper [10] presents a hybrid energy system combining a 114 kW solar PV array and a 30 kW PEM fuel cell. Solar energy powers the electrolyzer during the day to produce green hydrogen, which is stored and used by the fuel cell at night.

The feasibility of adopting DC systems in RTS is becoming evident, especially with the high penetration of DC-supplied loads, and the presence of power electronics technologies. The load supplied by Solar PV, can be directly fed to DC load, telecommunication tower, data centres, spacecraft and ship loads [11]. There is a possibility of direct charging of EV with PV, [12] proposed a battery-free DC microgrid with a distributed charging strategy to charge EVs solely by PV. This will reduce the payback period from 9.4 to 4.5 years, compared with a conventional solar charging roof top system. The paper [13] introduces a residential-scale demonstrator integrating a latent thermal energy storage (LTES) system with rooftop PV and heat pumps to increase self-consumption and energy efficiency in heating and cooling. Similarly, Shaikh et al. [14] demonstrate the utilization of a thermal storage unit used to provide space cooling in regions where the diurnal temperature range is favourable.

This study proposes an integrated approach for maximizing on-site utilization of surplus solar energy by addressing additional household loads, including water heating, space cooling, electric vehicle (EV) charging, and green hydrogen generation for cooking and future energy applications. Energy requirements for these loads were calculated considering a typical Indian household scenario. Results indicate that even after meeting the complete energy demands for heating, cooking, and EV charging, a considerable surplus remains, which can be fed back into the grid. This approach enhances the overall energy utilization of rooftop PV systems, improving economic returns and promoting sustainable energy integration. A detailed analysis is shown for the city Nagpur for one residential consumer load pattern with 5 kWp RTS installation in achieving carbon-neutral households. The objective is to design and evaluate a system that can meet the electrical, thermal, and cooking energy needs of a family of four to six while providing EV charging and a small unit of battery for emergency supply in the grid failure event.

Further, the same energy consumption pattern is compared for different geographical locations in India for the 5 kWp system.

Section 2 describes the detailed system architecture, specification of PV system is given in Sect. 3, and Sect. 4 gives the details of generation and utilization of energy in electrical appliances and other non-electrical loads like cooking and water heating. A detailed energy assessment for the RTS is given in Sect. 5. Comparison of energy utilization and surplus assessment in various Indian Cities is given in Sect. 6, followed by the conclusion in Sect. 7.

2 System Design

The proposed system given in the Fig. 1 integrates a rooftop Solar Photovoltaic system with various utilization pathways essential for any household. Typical pathways include thermal energy storage unit, electrolysis for hydrogen cooking, vehicle charging and emergency battery charging for emergency lighting load in case grid unavailability. Each unit shown in the diagram is connected to the DC side of the PV system, so that generated power can be directly utilized with less energy conversion stages to ensure reduced loss and increasing energy. This will also increase energy yield and thereby have maximum utilization of converted solar energy through PV systems. Each unit will require a dedicated dc-dc converter for working at appropriate voltage. And one common dc-dc converter for maintaining maximum power point tracking (MPPT). Centralized current control provides a reference power to each dc converter as per need and availability of the solar energy. Customers select the power setting of each unit through a centralized converter depending on the need and availability. The system can be designed for auto-operation to generate the power reference for each unit depending on optimization principle. IoT sensors will provide various data for optimization of the system.

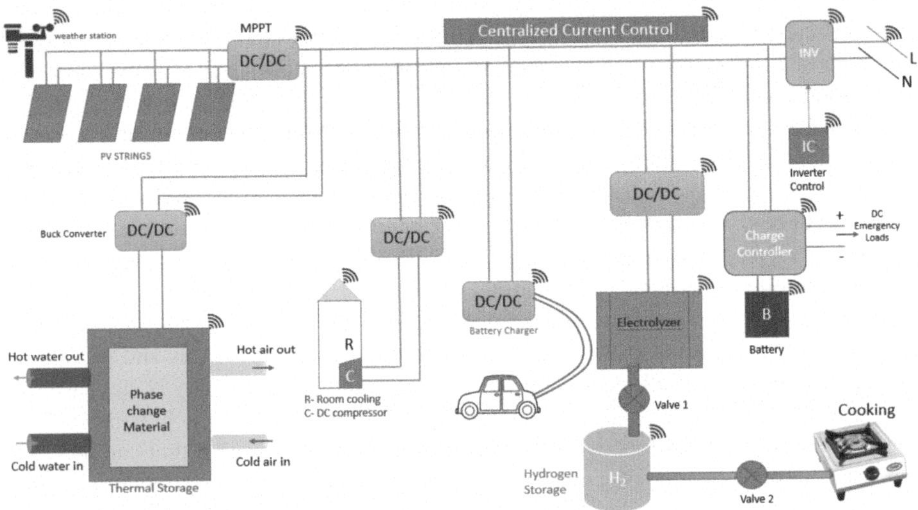

Fig. 1. PV System with Thermal Storage, EV Charging, Hydrogen Cooking, Refrigeration, and Battery Charging for DC load

3 PV System Specifications

Table 4 shows a typical household in Nagpur city (Latitude 21.17° N, Longitude 79.10° E) in India with sanctioned demand of 3kW whose electrical energy monthly consumption pattern for four to six family members. Family required one LPG cylinder in a month and around 50 L/day of hot water for bathing. Family maintains two electric vehicles. Emergency lighting system is also maintained for grid failure with a small unit with a charge controller. Further sections calculate the monthly energy requirement for each unit. Also, the 5kWp rooftop system is installed at the household.

The default value for the system losses is considered to be 15% adding losses of various categories mentioned in the Table 1 based on the PV watt calculator [15] provided by National Renewable Energy Laboratory.

Table 1. Default Loss Assumptions for PV System Performance

Category	Default Value (%)
Soiling	2
Shading	3
Snow	0
Mismatch	2
Wiring	2
Connections	0.5
Light-Induced Degradation	1.5
Nameplate Rating	1
Age	0
Availability	3
Total losses	15%

Table 2 denotes the PV system specification for 5kWp RTS installation at a particular household. A standard module used is of crystalline silicon type with efficiency of 19% [8].

Table 2. PV System Specification for Nagpur City [15]

DC System Size	5 kWp
Module Type	Standard
Array Type	Fixed (open rack)
System Losses	15%
Array Tilt	21°
Array Azimuth	180°
Inverter Efficiency	96%

The proposed system design integrates a rooftop Solar Photovoltaic (PV) system with multiple energy storage and utilization pathways to achieve maximum energy efficiency, sustainability, and grid independence. This hybrid configuration supports not only household energy needs but also addresses thermal storage, hydrogen generation, and electric vehicle (EV) charging requirements.

4 Generation and Utilization of Energy

Generation of the electrical energy is through the 5 kWp solar PV system. This energy can be converted into thermal energy for water heating, hydrogen energy for cooking and energy required for charging the vehicle. Generated energy varies with the season.

Nagpur city has three prominent seasons; summer season (February to June), rainy season (June to September), winter season (October to January). Accordingly, the output of the PV system varies. Utilization of electrical energy is more during April to June because of more cooling loads. Utilization of thermal energy will be different which is almost nil during summer season. Cooking energy requirement and energy requirement for electric vehicles is considered to be constant. It is assumed that, two electric scooters are maintained by the family. In the following section the calculations for energy generation and utilization is performed.

4.1 Photovoltaic (PV) Energy

At the core of the system are rooftop-mounted solar PV panels designed to harness solar radiation and convert it into direct current (DC) electricity. The panels are optimally oriented and tilted based on the site's geographic location to ensure maximum annual energy yield. Photovoltaic (PV) systems convert solar radiation directly into electrical energy using semiconductor materials. The amount of energy a PV system can generate depends on several factors, including the installed capacity of the system, the availability of sunlight at the location, and the overall efficiency of system components.

Electrical energy generation is calculated from the following equation,

$$W = P \times PSH \times Days \times PR \tag{1}$$

where,

W = total units generated
P = system size (5 kWp)
PSH = peak sun hours per day (6 h for 6 kWh/m^2-day average irradiation)
$Days$ = days in month (30)
PR = performance ratio (**0.75** for rooftop)
$= 5 \times 6 \times 30 \times 0.75$
= 675 kWh/month

Based on this monthly electrical unit generation for each month can be calculated and is given in Table 3.

The average sunlight hours per day is calculated using monthly energy output from a 5 kW solar PV system with a performance ratio of 0.75. The results show higher sunlight availability in March, April, and May (above 5.7 h/day), while July and August show the lowest values (around 3.2–3.5 h/day) due to monsoon conditions. These variations directly impact solar energy generation, highlighting the importance of incorporating storage systems to ensure a consistent energy supply throughout the year.

Table 3. Monthly Solar Energy Generation and Average Sunlight Hours for Nagpur City [15]

Month	Average Peak Sunlight Hours/Day	Energy Generated (kWh)
Jan	5.47	678
Feb	5.67	635

(continued)

Table 3. *(continued)*

Month	Average Peak Sunlight Hours/Day	Energy Generated (kWh)
Mar	5.90	732
Apr	5.80	696
May	5.76	714
June	4.34	521
July	3.55	440
Aug	3.23	401
Sep	4.38	525
Oct	5.27	653
Nov	5.78	693
Dec	5.45	676

4.2 Hydrogen Generation for Cooking

Cooking demand for a family of four to six members is one LPG cylinder (14.2 kg) per month, and calorific value of LPG is 12.78 kWh/kg.

Total cooking energy requirement for a month by the family = 181.47 kWh

Calorific value of hydrogen is 33.3 kWh/kg

Amount of hydrogen required for cooking in one month = 181.47/33.3 = 5.44 kg

For producing 1 kg of hydrogen from electrolyzer it consumes almost 50 kWh of electrical energy.

So, total electrical energy required for producing 5.44 kg of hydrogen = 272.48 kWh

4.3 Thermal Energy Storage

Assuming 50 L daily requirement of hot water by the family for bathing. It is required to raise water temperature up to 65 °C.

We use:

$$Q = m \times Cp \times \Delta T \tag{2}$$

where:

- Mass of water = 50 kg (1 L $\approx$ 1 kg)
- C_p = specific heat of water = 4.186 kJ/kg·K
- ΔT = temperature rise (assume heating from 25 °C $\rightarrow$ 65 °C)

$Q = 50 \times 4.186 \times 40 = 8,372$ kJ = 2.32 kWh.

So monthly energy requirement = $2.33 \times 30 = 70$ kWh/month.

However, the thermal energy demand varies throughout the year due to seasonal temperature fluctuations and changing user needs. Higher thermal energy requirements are observed during the colder months (e.g., January, June to December) for water and space heating. Typical summer Months (March, April, and May) require comparatively less thermal energy and it is assumed to be zero.

4.4 EV Battery Storage

It is assumed that a family uses two electric two-wheelers, with a combined monthly demand of 80 kWh, are charged using a lithium-ion battery system.

Energy Needed for 2 Electric Two-Wheelers.
One scooter battery: 3 kWh.
Full charge range: 100 km.
Monthly travel: 1,200 km (=40 km/day).
Energy per scooter:
$(1{,}200\,/100) \times 3$ kWh $= 36$ kWh.
For 2 scooters:
$36 \times 2 = 72$ kWh.

4.5 Emergency Battery Storage

Emergency battery storage is to be maintained for operating emergency DC lights during grid failure at night and few fans. It needs to maintain a battery with charge controller with one DC light in each room including wash rooms. As chances of grid failure at night time are low it is assumed that this will consume only 2 kWh each month.

Table 4. Monthly Allocation of Energy

Month	Energy generated per month from PV (kWh)	Electricity consumption per month (kWh)	Thermal storage per month (kWh)	Hydrogen production per month (kWh)	Vehicle Charging per month (kWh)	Emergency battery system (kWh)	Surplus Power per month (kWh)
Jan	678	107	70	273	72	2	154
Feb	635	117	70	273	72	2	101
Mar	732	126	10	273	72	2	249
Apr	696	190	10	273	72	2	149
May	714	225	5	273	72	2	137
June	521	210	40	273	72	2	−76
July	440	139	70	273	72	2	−116
Aug	401	132	70	273	72	2	−148
Sep	525	125	70	273	72	2	−17
Oct	653	143	70	273	72	2	93
Nov	693	122	70	273	72	2	154

(continued)

Table 4. (*continued*)

Month	Energy generated per month from PV (kWh)	Electricity consumption per month (kWh)	Thermal storage per month (kWh)	Hydrogen production per month (kWh)	Vehicle Charging per month (kWh)	Emergency battery system (kWh)	Surplus Power per month (kWh)
Dec	676	110	70	273	72	2	149
Total	7364	1746	625	3276	864	24	829

5 Energy Assessment

Table 4 presents a comprehensive overview of the monthly energy generation from a solar PV system and its distribution across various uses, such as household consumption, thermal storage, hydrogen cooking, charging EVs and emergency battery storage. The system generated a total of 7,364 kWh over the year, with production varying by month from a high of 732 kWh in March to a low of 401 kWh in August.

The generated energy is allocated monthly to:

- Household consumption**:** 107–225 kWh, highest in May.
- Thermal storage**:** 5–70 kWh.
- Hydrogen production**:** fixed at 273 kWh.
- Vehicle charging**:** fixed at 72 kWh.
- Emergency battery system**:** fixed at 2 kWh.

Surplus power varies between months, reaching a maximum of +249 kWh in March and falling into a deficit in June (–76 kWh), July (–116 kWh), and August (–148 kWh), indicating demand exceeded solar supply during these months. PV output power will also be reduced because of the cloudy sky during June to August. The annual surplus total 829 kWh, showing that overall generation exceeded allocated uses across the year, though seasonal shortages occur in mid-year.

The largest share of the solar energy goes toward hydrogen generation, reflecting a major focus on clean fuel production. With 257 kWh allocated each month, this accounts for a total of 3276 kWh annually. The household electricity usage consumes a moderate portion of the energy, totaling 1746 kWh/year. This includes lighting, appliances, and other domestic needs. A consistent 70 kWh/month is used for thermal applications such as water heating or space heating which is almost negligible during summer, making up 625 kWh annually. Energy used for electric vehicle charging (72 kWh/month, total 864 kWh/year) supports clean transport solutions, likely for 2 wheelers or a small EV fleet. Approximately 829 kWh of excess or surplus energy remains, which could be stored in additional batteries or exported to the grid or consumed with increasing electrical appliances like air conditioners for more comfort.

6 Energy Utilization and Surplus Assessment in Various Indian Cities

In this section a detailed analysis of a few more cities in India is considered to understand the surplus/deficit energy available after supplying the required loads. In doing so, the same pattern of electricity consumption, hydrogen requirement for cooking, EV charging and Emergency battery requirement is assumed. However, the thermal unit consumption is location specific and is different for different locations. It is calculated based on the average lowest temperature of the day [16]. Table 5 shows the location of different cities considered for analysis and their average annual solar radiation along with annual PV output for 5 kW system configuration given in Table 1 and Table 2.

Table 5. Average Solar Radiation and Annual PV Output for Different Cities in India [15]

Sr. No	City (State)	Latitude	Longitude	Average annual Solar radiation (kWh/m^2/Day)	Annual PV Output (kWh)
1	Bhavnagar (Gujarat)	21.7	72.1	5.96	7750
2	Chennai (Tamilnadu)	13.0	80.3	5.5	7364
3	Jammu(Jammu)	32.7	74.8	5.39	7082
4	Jodhpur(Rajasthan)	26.2	73.0	6.2	8064
5	Bengaluru(Karnataka)	13.2	77.7	5.73	7694
6	Leh(Ladakh)	34.1	77.5	6.34	9352
7	Nagpur(Maharashtra)	21.1	79.1	5.52	7285
8	Dwarka(Gujarat)	22.4	69.0	6.09	8220
9	Thiruv-Ananthapuram(Kerala)	8.4	76.9	5.11	6826
10	Wokha(Nagaland)	26.0	94.2	5.18	6806

Table 6, reports the annual electrical energy generation from the 5kWp RTS installation system and the annual unit requirement for the self-consumption in electrical appliances, thermal application, EV charging, hydrogen generation for cooking, and maintaining emergency battery supply. The same is graphically shown in Fig. 1. After utilization for the self-consumption the annual l surplus unit generation is transported to the grid. As observed from Table 6, almost, all the RTS installations in considered cities, are generating surplus except Wokha city located in Nagaland. This is because of higher annual thermal unit's requirement. Thermal unit requirement is also large for the cities like Leh and Jammu, but it is compensated by the larger unit of solar power generation. All three cities are situated in areas dominated by hilly or mountainous landscapes of Himalaya. Leh is a high-altitude cold desert plateau town located in Ladakh (Fig. 2).

Table 6. Annual Allocation of Energy for Various Indian Cities

City	Average Irradiance (kWh / m^2 / day)	Solar Unit Generation	Electrical Unit Consumption	Thermal Units (kWh)	EV Charing (kWh)	Units of for H_2 (kWh)	Battery Units (kWh)	Surplus Power (kWh
Bhavanagar	5.96	7750	1746	873	864	3276	24	966
Chennai	5.5	7364	1746	841	864	3276	24	612
Jammu	5.39	7082	1746	1048	864	3276	24	123
Jodhpur	6.2	8064	1746	935	864	3276	24	1218
Bangluru	5.73	7694	1746	974	864	3276	24	809
Leh	6.34	9352	1746	1623	864	3276	24	1818
Nagpur	5.52	7285	1746	903	864	3276	24	471
Dwarka	6.05	8222	1746	852	864	3276	24	1459
Thiruvananthapuram	5.11	6826	1746	866	864	3276	24	49
Wokha	5.18	6806	1746	1015	864	3276	24	-119

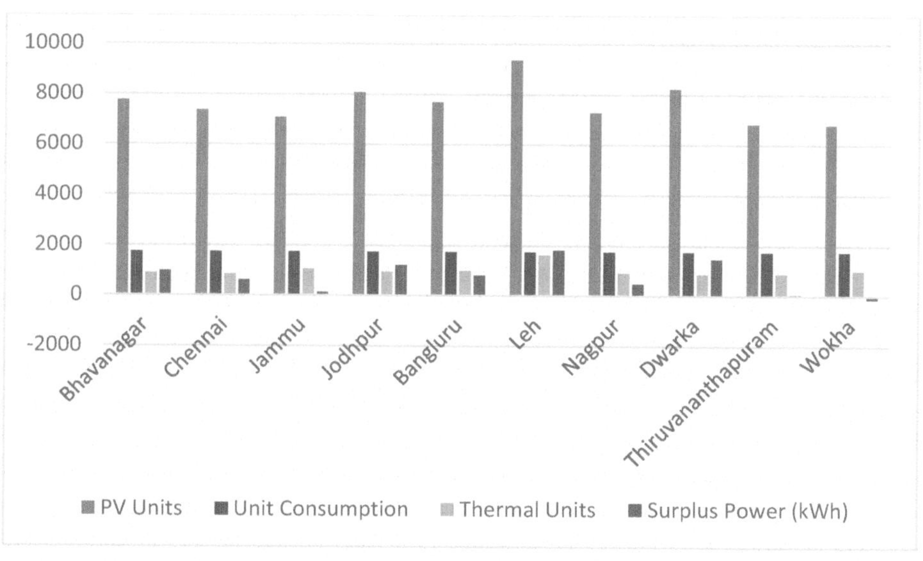

Fig. 2. Solar power generation, Electrical Unit Consumption, thermal units, surplus unit generation

7 Conclusion

The proposed system introduces an innovative, multi-vector energy solution integrating AC loads, DC loads, thermal energy storage, green hydrogen generation, and EV charging through rooftop solar PV. A household with 3 kW sanctioned demand connected with a 5 kW solar rooftop system for various cities in India completely satisfies the energy needs of the household as well as generating surplus power. This paper offers a replicable and future-ready solution to India's residential energy challenges. It ensures energy reliability, promotes clean alternatives, and provides a techno-economically viable model with strong potential for commercialization and public-sector adoption.

References

1. Parra, D., Patel, M.K.: The nature of combining energy storage applications for residential battery technology. Appl. Energy **239**(July 2018), 1343–1355 (2019)
2. Luthander, R., Widén, J., Nilsson, D., Palm, J.: Photovoltaic self-consumption in buildings: a review. Appl. Energy **142**, 80–94 (2015)
3. Gong, H., Rallabandi, V., Ionel, D.M., Colliver, D., Duerr, S., Ababei, C.: Dynamic modeling and optimal design for net zero energy houses including hybrid electric and thermal energy storage. IEEE Trans. Ind. Appl. **56**(4), 4102–4113 (2020)
4. Ntube, N., Li, H.: Optimal sizing of Battery Energy Storage System in a PV-Battery system for a Net Zero Home, IEEE PES Innovative Smart Grid Technologies Conference - Latin America (ISGT Latin America), pp. 1–5. Lima, Peru (2021)
5. Rasouli, V., Hemmati, R.: Net zero energy home including photovoltaic solar cells, wind turbines, battery energy storage systems and hydrogen vehicles. In: 2017 International Conference in Energy and Sustainability in Small Developing Economies (ES2DE), Funchal, Portugal, pp. 1–5 (2017)
6. Roy, A., Arun, P., Bandyopadhyay, S.: Design space for isolated power sys-tems-a deterministic approach. SESI J. **17**(1), 54–69 (2007)
7. Roy, A., Kedare, S.B., Bandyopadhyay, S.: Physical design space for Isolat-ed wind-battery system incorporating resource uncertainty. In: Proceedings of the Institution of Mechanical Engineers, Part A: Journal of Power and Energy, vol. 225(4), pp. 421–442 (2011)
8. Phap, V.M., et al.: Feasibility analysis of hydrogen production potential from rooftop solar power plant for industrial zones in Vietnam. Energy Rep. **8**, 14089–14101 (2022)
9. Topriska, V.: Maria Kolokotroni, Zahir Dehouche, Earle Wilson : solar hydrogen system for cooking applications: experimental and numerical study. Renewable Energy **83**, 717–728 (2015)
10. Tawil, H., Albarghot, M.M., Abraheem, E.B.: A sizing and dynamic model for a green hydrogen as energy storage technique for the hybrid system 50KW solar PV with PEM fuel cell. In: 14th International Renewable Energy Congress (IREC), pp. 1–6, Sousse, Tunisia (2023)
11. Elsayed, A.T., Mohamed, A.A., Mohammed, O.A.: DC microgrids and distribution systems: an overview. Electric Power Syst. Res. **119**, 407–417 (2015)
12. Liu, X., et al.: Charging private electric vehicles solely by photovoltaics: a battery-free direct-current microgrid with distributed charging strategy. Appl. Energy **341**, 121058 (2023)
13. Waser, R., Berger, M., Maranda, S., Worlitschek, J.: Residential-scale demonstrator for seasonal latent thermal energy storage for heating and cooling application with optimized PV self-consumption, IEEE PES Innovative Smart Grid Technologies Europe (ISGT-Europe), pp. 1–5, Bucharest, Romania (2019)
14. Roy, A., Shaikh, U., Kale, S., Sur, A.: Performance analysis of an energy-efficient PCM-based room cooling system. Front. Heat Mass Transfer (FHMT) 20 (2023)
15. NREL. PVWatts® Calculator. National Renewable Energy Laboratory, Golden, CO, USA. https://pvwatts.nrel.gov/. Accessed 15 Aug 2025
16. https://en.climate-data.org/

Automated Assessment of Rooftop Solar Energy Potential Using Deep Learning-Driven Building Footprint Extraction Model

Sachin Mishra(✉)

International Institute of Information Technology Bangalore, Bangalore, India
sachin.mishra@iiitb.ac.in

Abstract. Accurate rooftop segmentation from high-resolution satellite images is crucial for assessing urban solar energy potential on a larger scale. Traditional methods, such as U-Net and Mask R-CNN architectures, have proven to be effective in general semantic segmentation tasks. However, they typically exhibit limited integration with solar irradiance modeling and frequently misclassify complex urban scenes due to shadows, building occlusions, and heterogeneous rooftop objects. In this work, we propose a specialized deep learning architecture that combines a ResNet-34 encoder pretrained on ImageNet and fitted to three-channel RGB satellite inputs with a custom U-Net decoder that is enhanced with attention gates to selectively emphasize boundary details. Our model significantly outperforms previous rooftop detection networks, achieving an Intersection over Union (IoU) score of 0.9575 on a multi-source annotated rooftop dataset. We integrate the extracted rooftop footprints with geographic solar irradiance maps to provide high-resolution solar potential calculations. This converts segmentation outputs into actionable energy insights. The segmentation reliability and the end-to-end energy modeling are validated by a 15âĂŞ30% improvement in estimation accuracy when compared to typical GIS-based solar evaluation pipelines. Our framework provides a scalable and reliable tool for stakeholders in renewable energy and urban planning, enabling accurate identification of rooftop locations and data-driven decision-making for deploying distributed solar systems.

Keywords: Solar potential · Mask R-CNN · ImageNet · ResNet34 · U-Net · Attention gates · Encoder · Decoder · rooftop footprint · Rooftop segmentation

1 Introduction

The global energy demand is escalating at an unprecedented rate, driven by population growth, urbanization, and industrialization. Concurrently, climate change has intensified the need for sustainable energy solutions [1–3], with solar

© The Author(s), under exclusive license to Springer Nature Switzerland AG 2026
A. Kannan et al. (Eds.): ADCOM 2025, CCIS 2947, pp. 99–114, 2026.
https://doi.org/10.1007/978-3-032-26269-1_7

energy emerging as a pivotal component of the renewable energy transition. Among solar energy technologies [4,5], rooftop solar systems provides a decentralized and scalable solution [6], particularly in urban areas where land availability is limited. However, the widespread adoption of rooftop solar energy remains hindered by the lack of efficient and scalable methods for assessing solar potential.

Traditional approaches, such as GIS-based manual assessments [7–9], are time consuming, labor-intensive, and often lack the precision required for large-scale deployment. These methods struggle to account for inconsistent building footprints and shading effects, leading to inaccurate estimations of solar energy potential. Moreover, the absence of standardized tools for integrating building footprint extraction with solar irradiance data further exacerbates the challenge, particularly in developing regions like India, where rapid urbanization demands cost-effective and scalable solutions.

Recent advancements in deep learning (DL) have demonstrated remarkable success in automating complex tasks such as image segmentation and object detection [10,11]. However, the integration of DL for rooftop segmentation and solar potential estimation remains underexplored, especially in resource-constrained settings. Existing studies either focus on high-cost LiDAR-based methods [12,13] or lack the scalability to handle diverse urban landscapes. This gap underscores the need for a scalable and automated pipeline that combines accurate rooftop extraction with localized solar irradiance data to generate reliable solar potential maps.

This study addresses these challenges by proposing a deep learning-driven framework for rooftop segmentation and solar energy estimation. Leveraging state-of-the-art U-Net architectures [15] with attention mechanisms, the framework automates the extraction of building footprints from high-resolution satellite imagery. The extracted rooftops are then integrated with NASA POWER irradiance data to calculate solar energy potential [17], enabling the generation of open-access solar potential maps. By bridging the gap between DL-based rooftop segmentation and solar energy modeling, this work provides urban planners and policymakers with a scalable tool for identifying viable regions for solar installations, thereby accelerating the transition to sustainable energy systems. The prevalent acceptance of CNNs is owed to the ingenious methods devised to facilitate their training, wherein these networks have now become an indispensable tool in visual data analysis. The transformative potential of deep networks, particularly in modeling intricate response functions with data, has further improved their popularity in the broader machine learning domain.

1.1 Motivation

With the rising demand for renewable energy sources [18] and the global emphasis on sustainable urban development, efficient rooftop detection and building footprint extraction from satellite [19–21] and aerial imagery have become crucial components of geospatial analysis. One of the key applications is in assessing solar energy potential, where accurate identification of rooftop areas enables informed decisions on solar panel installations [22–24]. However, traditional methods relying on manual mapping or classical computer vision techniques often struggle with occlusions, varying roof types, shadows, and inconsistent image quality. The proposed work aims to address these challenges by leveraging deep learningâĂŤspecifically, semantic segmentation using U-NetâĂŤto automatically and accurately extract rooftop features from imagery. This automation not only accelerates the mapping process but also enhances accuracy, scalability, and adaptability across diverse urban and rural environments. The primary goal is to develop a robust segmentation model that can generalize across datasets and serve as a backbone for applications in smart cities, energy auditing, and geographic information systems (GIS).

1.2 Objective

The primary objective of this work is to design an accurate and automated rooftop segmentation model that can be applied to satellite and aerial imagery for urban planning and solar energy estimation. In many developing regions, there is a growing demand for tools that can help identify viable rooftop areas for solar panel installation, yet reliable datasets and methods to support such initiatives are limited. This work aims to bridge that gap by utilizing deep learning, particularly a U-Net-based semantic segmentation model, to extract rooftop footprints from annotated image data. To achieve this, multiple open-source annotated datasets in COCO format were merged and converted into a unified training structure, followed by the generation of binary masks to train the segmentation model effectively. The methodological data flow diagram of the proposed solution is represented in Fig. 1. Through this approach, this work not only contributes to the field of geospatial intelligence but also provides a scalable pipeline that can be integrated into broader systems for environmental monitoring, renewable energy adoption, and smart city development.

1.3 Contribution

The key contributions of this work are:

1. **Dataset Integration and Preprocessing:** Merged and standardized multiple datasets containing rooftop annotations in COCO format. Converted them into a format suitable for semantic segmentation tasks.
2. **Mask Generation Pipeline:** Developed a Python-based utility to extract binary segmentation masks from COCO annotations, automating the preprocessing pipeline.
3. **Model Design and Training:** Implemented and trained a U-Net-based deep learning model tailored for rooftop segmentation with customized input resolution and optimizer settings.
4. **End-to-End Workflow:** Developed a complete end-to-end framework-from raw annotated data to prediction outpu-demonstrating the potential for real-world applications such as solar rooftop mapping and smart city planning.

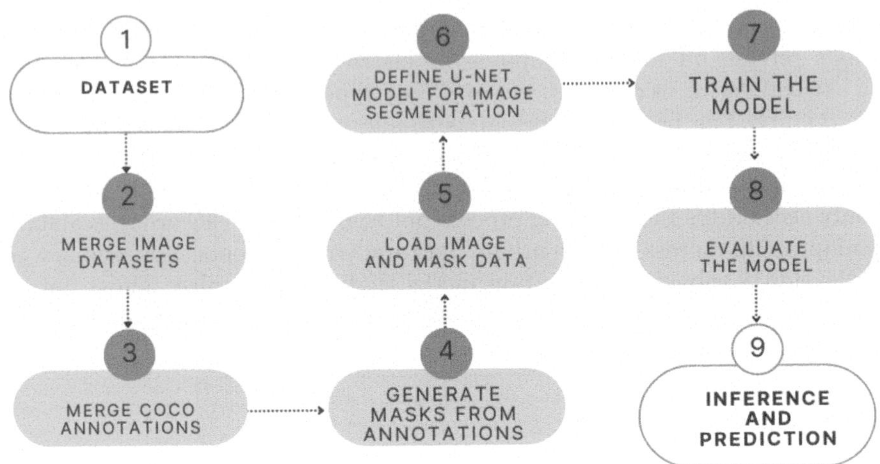

Fig. 1. Data Flow Diagram.

The rest of the paper is organized as follows: Sect. 2 outlines the literature review. Section 3 describes datasets used with sources and statistics. Section 4 discusses the proposed Methodology describing the architectural explanation of proposed solutions, and Sect. 5 presents the results of the proposed solution. Finally, Sect. 6 provides conclusions and future work.

2 Related Work

This section provides a comprehensive review of existing works on rooftop segmentation and solar potential estimation. We organize the prior works into two main categories: building footprint extraction techniques that leverage deep learning for automated rooftop detection and solar potential estimation methods that quantify energy generation capacity.

2.1 Building Footprint Extraction:

Ronneberger et al. [25] pioneered U-Net for biomedical segmentation, later adapted for rooftop extraction due to its encoder-decoder architecture. Zhang et al. [26] deployed YOLOv5 for rural rooftop detection but ignored shading effects, limiting urban applicability.

Mengge Chen et al. [27] applied Mask R-CNN on post-earthquake aerial imagery (Christchurch dataset) to detect damaged rooftops, achieving 79% accuracy. Limitations: No energy estimation; unclear dataset preprocessing. Roberto Castello et al. [28] Existing solar panels were detected with 94% accuracy using U-Net on Swiss topographic data. Limitation: Focused on panel recognition, not potential rooftops. Qingyu Li et al. [29] proposed AFM-based footprint generation (ISPRS dataset) with optimized DL models. Limitation: Struggled with blurred rooftops in low-resolution imagery.

2.2 Hybrid Approaches for Solar Potential Estimation

Laura Grunwald et al. [30] proposed GIS-based analysis of green roofs in Braunschweig, Germany. Limitations: No accuracy metrics; dataset not disclosed.

Singh et al. [31] combined CNN with GIS for flat-roof solar mapping but excluded sloped roofs. Roofpedia [32] mapped solar/green roofs across 17 cities using OpenStreetMap and ResNet50 (87.59% accuracy). Limitations: Focused on existing panels; no energy output estimates.

Zhaojian Huang et al. [33] U-Net-based solar potential mapping in Wuhan using Inria dataset (90.49% accuracy). Limitation: Analyzed building coverage, not rooftop-specific areas. Hasan Nasrallah et al. [34] proposed Multiclass U-Net for Lebanese rooftops (84.3% accuracy) with PV placement simulation. Limitation: Private dataset; region-specific. Taskin Jama et al. [35] proposed GIS/ArcGIS-based rooftop area mapping in Dhaka. Limitations: No accuracy metrics; dataset not described.

3 Dataset Description

This section presents detailed information about the dataset used for training and evaluating our rooftop segmentation model. The section describes the multi-source data collection process, including high-resolution satellite images from Google Earth. Dataset statistics and usage details are also presented.

3.1 Data Sources

The datasets were gathered from GIS, Microsoft Planet, Google Earth, Nearmap platforms; the sample images are represented in Fig. 2. The dataset was split into three parts for training, testing, and validation, as represented under the Dataset Statistics subsection in Table 1. The datasets were stored in a drive in different folders and subsequently merged for efficient working. The polygon masks for the data are being used by Microsoft Planetary. The rooftops are not hidden in any of the other two sources. Therefore, the data was annotated, and masks were generated using OpenCV's library & generate-masks() function written in Python for making the dataset.

Table 1. Dataset splits, image formats, and annotation types.

Split	No. of Images	Format	Annotation Type
Train	1494	.jpg/.png	COCO JSON
Validation	280	.jpg/.png	COCO JSON
Test	221	.jpg/.png	COCO JSON

Fig. 2. Sample dataset, collected using google earth and nearmap APIs.

3.2 Dataset Statistics

3.3 Description of Data and Its Usage

The study utilized high-resolution satellite imagery from Google Earth and the National Agriculture Imagery Program (NAIP), offering 1-meter spatial resolution. These datasets were selected for their accessibility, coverage of urban and peri-urban regions, and compatibility with building footprint extraction tasks. Images were captured in a variety of locations, times of year, and lighting scenarios to ensure diversity. RoboFlow, a collaborative platform for curating labeled datasets, provides the annotations in the COCO (Common Objects in Context) format through crowdsourcing them. The COCO format provided structured metadata, including polygon coordinates for rooftops, enabling seamless integration with deep learning frameworks.

4 Methodology

This section presents our comprehensive framework for automated rooftop segmentation and solar potential assessment. The complete pipeline into four key components: data preprocessing, feature extraction, proposed model architecture, and finally, rooftop area calculation with energy estimation.

4.1 Data Processing

All datasets were:

1. Unzipped and stored in a structured directory format.
2. Merged by combining image files and merging their corresponding COCO annotation JSONs.

3. Converted into binary segmentation masks using a custom generate-masks() function that transformed COCO polygons into black-and-white masks, where white regions represent rooftops.

Image and Mask Specifications

1. **Input Image Size:** All images and masks were resized to a fixed dimension of 128x128 pixels to ensure uniformity for training the deep learning model.
2. **Mask Format:** Binary masks (single-channel greyscale images) were generated for training the segmentation model. Pixel values are either 0 (background) or 1 (rooftop).

4.2 Feature Extraction

In this work, the primary form of feature extraction is performed automatically through the convolutional layers of the U-Net architecture, eliminating the need for manual or handcrafted feature engineering.
Input Features: Each input image is a satellite or aerial view, resized to 128×128 pixels with 3 RGB channels. These images serve as raw input features containing: color intensity patterns, texture information, Shadows & illumination variations, and Structural outlines (e.g., edges of roofs and buildings).
Learned Features: The U-Net architecture extracts hierarchical and spatially rich features at different depths through its encoder-decoder structure:

1. **Low-level features:** Captured in early convolutional layers, focusing on edges, colors, and small patterns.
2. **Mid-level features:** Extract structural shapes and textures such as repetitive rooftop tiles, boundaries, and shadows.
3. **High-level features:** Aggregate contextual information and distinguish between rooftop and non-rooftop regions.

Skip connections in U-Net help in preserving fine-grained spatial details, which is especially important for segmenting small or irregularly shaped rooftops.
Output Feature: The final output is a binary segmentation mask (1 channel) where:

1. Pixel value = 1 denotes rooftop region.
2. Pixel value = 0 denotes background or non-rooftop.

This mask is later used to calculate rooftop area and can be combined with geospatial coordinates for solar potential mapping.

4.3 Proposed Methodology

The model formation methodology involves implementing a U-Net architecture for rooftop segmentation. The U-Net consists of an encoder, which extracts hierarchical features using convolutional and pooling layers; a bottleneck for feature compression; and a decoder that restores spatial resolution through upsampling layers, the representation of the generalized architecture of the U-Net model with Attention (AAM) architecture is shown in Fig. 3. The model is compiled with the Adam optimizer and binary cross-entropy loss to optimize pixel-wise classification. For effective training, the input images and masks are normalized and resized to 128×128 pixels. Accuracy and loss metrics are used to evaluate the network's performance after it has been trained on the prepared dataset. Finally, IoU scores and visual comparisons of predicted masks are used to assess the efficacy of the trained model.

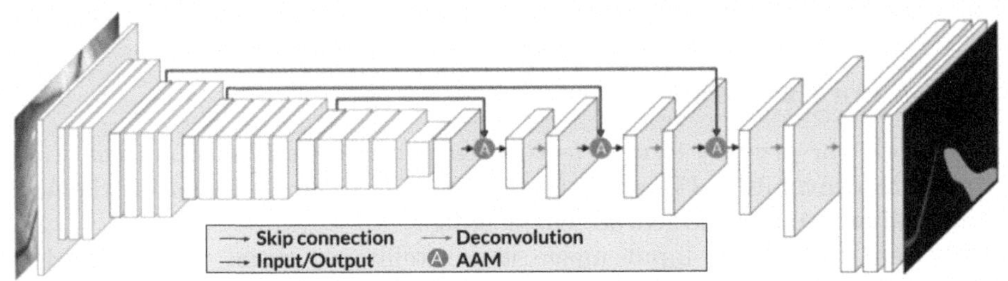

Fig. 3. UNET model with Attention (AAM)

Network Design: The proposed framework employs a U-Net architecture with enhancements to optimize rooftop segmentation accuracy and computational efficiency. The model combines a robust encoder for feature extraction with attention-guided decoding for precise boundary delineation; this network design is explained in Fig. 4.

Encoder Using ResNet-34 Backbone: The encoder leverages a ResNet-34 backbone pretrained on ImageNet, modified for satellite imagery:

1. Feature Extraction: The pretrained ResNet-34 extracts hierarchical features through five convolutional blocks, capturing multi-scale rooftop patterns.
2. Adaptations:
   (a) Input Channels: The first convolutional layer was modified to accept 3-channel RGB inputs (original: 3-channel natural images).
   (b) Skip Connections: Intermediate feature maps from ResNet stages 2âĂŞ5 are preserved for decoder fusion.
3. Transfer Learning Benefits: Pretrained weights accelerate convergence and improve generalization on limited rooftop datasets.

The ResNet-34 provides the optimal balance between accuracy, efficiency, and transfer learning. The skip connections in ResNet architecture are particularly beneficial for preserving fine-grained spatial details critical for accurate roof boundary segmentation. **Decoder: Attention-Guided Upsampling:** The decoder refines features through a series of up-sampling blocks with attention gates to prioritize rooftop boundaries.

1. Upsampling: Transposed convolutions (stride = 2) restore spatial resolution.

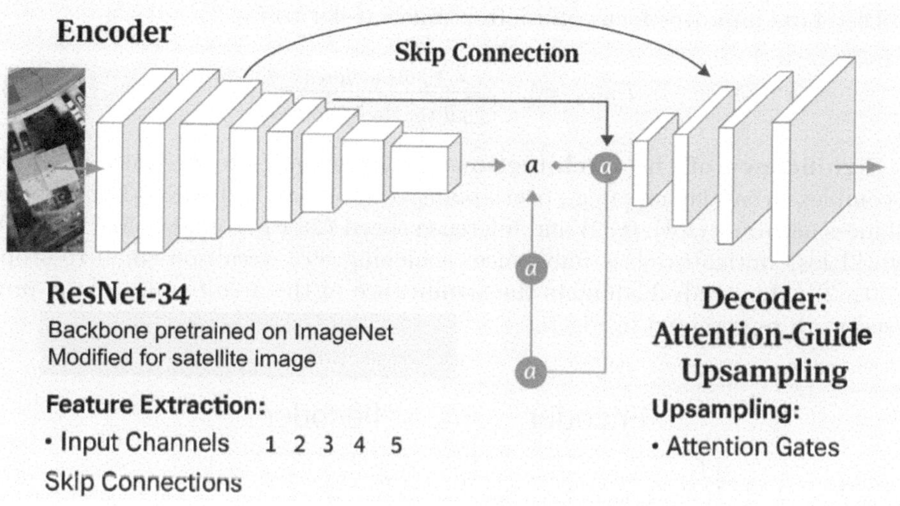

Fig. 4. Decoder Architecture.

2. Attention Gates Mechanism: Gates compute pixel-wise attention scores from encoder and decoder features. These gates suppress irrelevant background regions while emphasizing rooftop boundaries.

$$\alpha_{ij} = \sigma\left(W_g\, g_{ij} + W_x\, x_{ij} + b\right) \tag{1}$$

The attention coefficient is denoted by $\alpha_{i,j}$, the sigmoid activation by σ, b is the bias, and the encoder and decoder features, denoted by $x_{i,j}$ and, $g_{i,j}$ respectively, are denoted in Equation (1).
Attention gates function as soft feature selectors that learn to suppress irrelevant background regions while amplifying rooftop features. At each decoder level, the attention coefficient α_{ij} (Eq. 1) is computed by:
   (a) Query $g_{i,j}$: Coarse features from decoder indicating 'where to look.'
   (b) Key $x_{i,j}$: Fine features from encoder providing 'what is there.'
   (c) Gate activation: Element-wise addition followed by sigmoid, producing attention weights $\in [0, 1]$. These weights modulate the encoder features before concatenation with decoder features, effectively creating a learned skip connection.
3. Feature Fusion: Attended encoder features are concatenated with decoder outputs to recover fine-grained spatial details.

Loss Function: To address class imbalance between rooftops and background regions,a hybrid loss combining Dice Loss and Binary Cross-Entropy (BCE) is adopted. Dice loss directly optimizes the IoU metric and handles class imbalance by focusing on overlap between prediction and ground truth. BCE provides pixel-wise classification gradients, ensuring proper boundary delineation. The hybrid loss is calculated using Eq. (2):

$$\mathcal{L} = \lambda \cdot \mathcal{L}_{\text{Dice}} + (1 - \lambda) \cdot \mathcal{L}_{\text{BCE}} \tag{2}$$

Dice Loss improves focus on rooftop pixels (refer to Eq. 3):

$$\mathcal{L}_{\text{Dice}} = 1 - \frac{2 \sum y_{\text{true}} \, y_{\text{pred}}}{\sum y_{\text{true}} + \sum y_{\text{pred}}} \tag{3}$$

Significance of the Architecture: Attention gates reduce false positives in complex urban layouts (e.g., overlapping rooftops). The ResNet-34 backbone balances accuracy (IoU: 0.82) and inference speed (28 FPS on 512×512 images). Hybrid loss mitigates class imbalance, achieving 89% recall on small rooftops (<50 m2). This study highlights the significance of the architecture of the proposed work represented in Fig. 5.

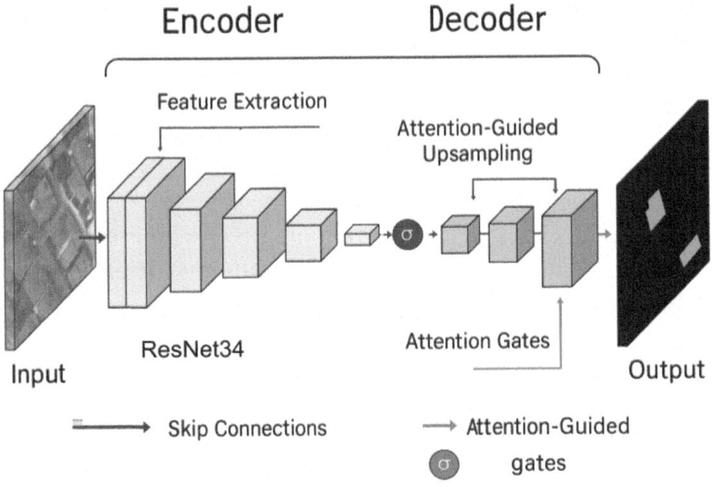

Fig. 5. Inner layers of Proposed Model Architecture.

4.4 Rooftop Area Calculation and Energy Estimation

Area Calculation: The model estimates rooftop area based on two features: size of the image and scale of measurement. The area in pixels is converted to square meters using the following Eq. (4):

$$\text{Area (m}^2) = \text{Area of image} \times \left(\frac{\text{m}}{\text{pixel}}\right)^2 \tag{4}$$

The area in square meters is converted to feet by the following Eq. (5):

$$\text{Area (ft}^2) = \text{Area (m}^2) \times 10.764 \tag{5}$$

Panel Coverage Estimation: Assuming a standard panel size for residential areas is 3.25×5.5 ft, the number of panels is estimated in Eq. (6):

$$\text{No. of panels} = \frac{\text{Usable rooftop area}}{\text{Size of panel}} \tag{6}$$

Solar Capacity: Each panel produces peak energy of 325 W (Wp), i.e., 0.325 kWp. Total Solar capacity is calculated in Eq. (7):

$$\text{Total Capacity} = \text{Number of Panels} \times 0.325 \tag{7}$$

Annual Energy Generation: It is calculated by the following Eq. (8):

$$\text{Annual Energy (in kW)} = \text{Total Capacity} \times \text{Derate Factor} \times \text{Equivalent Sun Hours} \tag{8}$$

where total capacity is total solar capacity produced by the building area, and the derate factor is 0.14 which accounts for the efficiency of the panel based on temperature, performance, and shading effect. Equivalent sun hours is the average number of hours of sun a particular location receives throughout the year based on the weather conditions.

5 Results

This section presents experimental results showing the effectiveness of our proposed framework. We evaluate model performance using standard segmentation metrics, including Intersection over Union (IoU), precision, recall, and loss curves throughout the training process. The results show that our attention-enhanced U-Net architecture achieves state-of-the-art performance, significantly outperforming existing rooftop detection methods. Detailed comparisons and performance analysis are presented in the following subsections.

5.1 Model Evaluation

The model is evaluated using Intersection over Union (IoU), confusion matrix, and loss metrics. Training, validation, and testing are conducted on a multi-source annotated dataset. The evaluation demonstrates the modelâĂŹs robustness and accuracy. Figure 6 illustrates the training and validation performance in terms of accuracy and loss throughout the learning process.

5.2 Model Performance

The accuracy of our proposed model and previously developed solutions are compared above. Table 2 displays a comparison of the suggested model with current rooftop detection models. The suggested U-Net model has an IoU of 95.75%, which is significantly higher than that of the current models. The attention gates of the decoder improve the accuracy of the model, resulting in better performance.

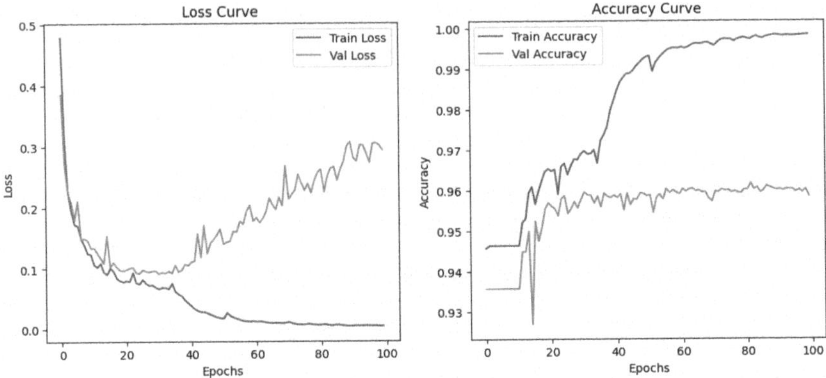

Fig. 6. Training and validation performance curves over 40 epochs. The model achieves convergence around epoch 30 with training accuracy of 97.2% and a validation accuracy of 95.8%. The minimal gap between training and validation curves indicates good generalization without overfitting. Early stopping was applied at epoch 40 when validation loss plateaued. The slight oscillation in validation metrics reflects the diversity of rooftop structures in the test set.

Table 2. Performance comparison of the proposed solution with previous work published on rooftop segmentation methods. Our proposed attention-enhanced U-Net achieves 95.75% accuracy, representing a 3.75% point improvement over the previous best (Castello et al., 92%)

Model	Dataset	Accuracy
Mask R-CNN [27]	Christchurch Dataset	79%
Multiclass U-Net (Lebanon) [34]	Private TUM Dataset	84.30%
ResNet50 (Roofpedia) [32]	OpenStreetMap + Mapbox API	87.59%
U-Net (Huang et al.) [33]	Inria Aerial Dataset	90.49%
U-Net (Castello et al.) [28]	Swiss Topographic Imagery	92%
Proposed U-Net	**RoboFlow (Custom)**	**95.75%**

Our proposed attention-enhanced U-Net achieves 95.75% IoU, representing a 3.75% point improvement over the previous best (Castello et al., 92%). Note that direct comparison is approximate, as methods were evaluated on different datasets with varying complexity. Our RoboFlow dataset includes challenging urban scenarios with heavy shadows and occlusions.

6 Conclusion and Future Work

This study demonstrates that a specialized deep learning pipeline combining an attention-guided U-Net decoder with a ResNet-34 encoder can achieve highly accurate rooftop segmentation from complex urban satellite imagery. Our model outperforms existing rooftop detection networks with an Intersection over Union (IoU) of 0.9575 by using pixel-wise attention gates to minimize background noise and skip connections to preserve multi-scale feature representations (Table 2). In comparison to traditional GIS-based procedures, the segmented rooftop footprints produced high-resolution potential maps that increased estimation accuracy by 15% to 30% when paired with geospatial solar irradiance data. These improvements highlight how useful it is to combine sophisticated convolutional feature extractors with specialized decoder attention mechanisms for applications using renewable energy. Our framework presents several practical advantages. First, it can be deployed in diverse metropolitan settings without requiring specialized hardware. Second, it relies on publicly available satellite and irradiance data, enhancing scalability. Third, it bridges the gap between semantic segmentation performance and practical energy insights, providing solar developers, policymakers, and urban planners with a robust automated tool for identifying and prioritizing rooftop locations for photovoltaic installations. Future research directions include real-time model deployment on streaming satellite feeds and integration of building height and roof tilt data to improve energy yield forecasts. These enhancements will further strengthen the framework's utility for large-scale solar potential assessment.

6.1 Generalization and Limitations

While our model achieves high performance on the RoboFlow dataset (predominantly Indian urban areas), several factors affect generalizability:

1 . Transferability Considerations: The attention-enhanced U-Net architecture with ResNet-34 encoder utilizes general-purpose features from ImageNet pretraining, suggesting good transfer potential to other geographic regions. Training data includes varied lighting conditions, seasons, and building types, though predominantly from urban areas.

2. Limitations:
   (a) Rural areas with non-standard roof materials (thatched, corrugated iron) may require fine-tuning on region-specific datasets.
   (b) Extreme weather conditions (e.g., heavy snow cover) are not represented in the training data and may affect performance.
   (c) Very low-resolution imagery ($\succ 2m/pixel$) may degrade segmentation accuracy.

6.2 Future Scope

This work lays a foundational framework for automated rooftop segmentation with significant potential for extensions and enhancements:

1. The model can be scaled to support high-resolution satellite imagery across diverse geographic regions, including rural and densely urbanized areas with complex rooftop structures.

2. Additionally, incorporating instance segmentation or object detection techniques could enable the identification of obstacles like water tanks, AC units, or chimneys on rooftops, further refining the estimations of usable space.
3. Future enhancements may include the use of more advanced deep learning architectures, such as Transformer-based models, or the incorporation of attention mechanisms to improve segmentation accuracy.
4. Lastly, deploying the system as a web-based interactive tool or mobile application can democratize access to rooftop analytics, empowering local governments, energy companies, and citizens to make data-driven decisions toward sustainable energy planning.
5. Integrate LiDAR for 3D rooftop modeling. Also incorporate DSM models to improve the estimation of angled roofs.

The potential for continued research and innovation in this domain remains vast, with promising opportunities to contribute to the development of more accurate, reliable, and accessible solutions for sustainable energy deployment.

References

1. Olabi, A.G., Abdelkareem, M.A.: Renewable energy and climate change. Renew. Sustain. Energy Rev. **158**, 112111 (2022)
2. Phebe Asantewaa Owusu and Samuel Asumadu-Sarkodie: A review of renewable energy sources, sustainability issues and climate change mitigation. Cogent Eng. **3**(1), 1167990 (2016)
3. Jones, D.W.: How urbanization affects energy-use in developing countries. Energy Policy **19**(7), 621–630 (1991)
4. Chu, Y., Meisen, P.: Review and comparison of different solar energy technologies. Global Energy Network Institute (GENI) **1**, 1–52 (2011)
5. Maka, A.O., Alabid, J.M.: Solar energy technology and its roles in sustainable development. Clean Energy **6**(3), 476–483 (2022)
6. Behura, A.K., Kumar, A., Rajak, D.K., Pruncu, C.I., Lamberti, L.: Towards better performances for a novel rooftop solar pv system. Sol. Energy **216**, 518–529 (2021)
7. Choi, Y., Suh, J., Kim, S.-M.: Gis-based solar radiation mapping, site evaluation, and potential assessment: a review. Appl. Sci. **9**(9), 1960 (2019)
8. Benalcazar, P., Komorowska, A., Kamiński, J.: A gis-based method for assessing the economics of utility-scale photovoltaic systems. Appl. Energy **353**, 122044 (2024)
9. Fahd Amjad and Liaqat Ali Shah: Identification and assessment of sites for solar farms development using gis and density based clustering technique-a case of pakistan. Renewable Energy **155**, 761–769 (2020)
10. Shaik, A., Balasundaram, A., Kakarla, L.S., Murugan, N.: Deep learning-based detection and segmentation of damage in solar panels. Automation **5**(2), 128–150 (2024)
11. Hoeser, T., Kuenzer, C.: Object detection and image segmentation with deep learning on earth observation data: a review-part i: evolution and recent trends. Remote Sensing **12**(10), 1667 (2020)
12. Suomalainen, K., Wang, V., Sharp, B.: Rooftop solar potential based on lidar data: Bottom-up assessment at neighbourhood level. Renewable Energy **111**, 463–475 (2017)

13. Tiwari, A., Meir, I.A., Karnieli, A.: Object-based image procedures for assessing the solar energy photovoltaic potential of heterogeneous rooftops using airborne lidar and orthophoto. Remote Sensing **12**(2), 223 (2020)
14. Li, C., et al.: Anu-net: attention-based nested u-net to exploit full resolution features for medical image segmentation. Comput. Graph. **90**, 11–20 (2020)
15. Alom, M.Z., Yakopcic, C., Hasan, M., Taha, T.M., Asari, V.K.: Recurrent residual u-net for medical image segmentation. J. Med. Imaging **6**(1), 014006 (2019)
16. Piragnolo, M., Masiero, A., Fissore, F., Pirotti, F.: Solar irradiance modelling with NASA WW GIS environment. ISPRS Int. J. Geo Inf. **4**(2), 711–724 (2015)
17. NASA Goddard Space Flight Center. Total and spectral solar irradiance sensor (TSIS-1). NASA Science (2024). https://science.nasa.gov/mission/tsis-1/
18. Nematollahi, O., Hoghooghi, H., Rasti, M., Sedaghat, A.: Energy demands and renewable energy resources in the Middle East. Renew. Sustain. Energy Rev. **54**, 1172–1181 (2016)
19. Kim, J., Bae, H., Kang, H., Lee, S.G.: CNN algorithm for roof detection and material classification in satellite images. Electronics **10**(13), 1592 (2021)
20. Song, Z., Pan, C., Yang, Q., Li, F., Li, W.: Building roof detection from a single high-resolution satellite image in dense urban area. In: Proceedings of the ISPRS Congress, pp. 271–277 (2008)
21. Park, J., Park, S., Kang, J.: Detecting and classifying rooftops with a CNN-based remote-sensing method for urban area cool roof application. Energy Rep. **11**, 2516–2525 (2024)
22. Malof, J.M., Hou, R., Collins, L.M., Bradbury, K., Newell, R.: Automatic solar photovoltaic panel detection in satellite imagery. In: 2015 International Conference on Renewable Energy Research and Applications (ICRERA), pp. 1428–1431. IEEE (2015)
23. Kumar, A.: Solar potential analysis of rooftops using satellite imagery. arXiv preprint arXiv:1812.11606 (2018)
24. House, D., Lech, M., Stolar, M.: Using deep learning to identify potential roof spaces for solar panels. In: 2018 12th International Conference on Signal Processing and Communication Systems (ICSPCS), pages 1–6. IEEE (2018)
25. Ronneberger, O., Fischer, P., Brox, T.: U-Net: convolutional networks for biomedical image segmentation. In: Navab, N., Hornegger, J., Wells, W.M., Frangi, A.F. (eds.) MICCAI 2015. LNCS, vol. 9351, pp. 234–241. Springer, Cham (2015). https://doi.org/10.1007/978-3-319-24574-4_28
26. Ding, W., Zhang, L.: Building detection in remote sensing image based on improved YOLOv5. In: 2021 17th International Conference on Computational Intelligence and Security (CIS), pages 133–136. IEEE (2021)
27. Chen, M., Li, J.: Deep convolutional neural network application on rooftop detection for aerial image. arXiv preprint arXiv:1910.13509 (2019)
28. Castello, R., Roquette, S., Esguerra, M., Guerra, A., Scartezzini, J.L.: Deep learning in the built environment: Automatic detection of rooftop solar panels using Convolutional Neural Networks. J. Phys. Conf. Ser. **1343**, 012034 (2019)
29. Liu, Z., Tang, H., Huang, W.: Building outline delineation from VHR remote sensing images using the convolutional recurrent neural network embedded with line segment information. IEEE Trans. Geosci. Remote Sens. **60**, 1–13 (2022)
30. Grunwald, L., Heusinger, J., Weber, S.: A GIS-based mapping methodology of urban green roof ecosystem services applied to a Central European city. Urban Forestry & Urban Greening **22**, 54–63 (2017)

31. Singh, A.: Assessment of potential rooftop solar PV electricity at a suburban scale, and a comparative analysis based on topographical obstruction and seasonality. PhD thesis, University of Wollongong (2023)
32. Abraham Noah Wu and Filip Biljecki: Roofpedia: automatic mapping of green and solar roofs for an open roofscape registry and evaluation of urban sustainability. Landsc. Urban Plan. **214**, 104167 (2021)
33. Huang, Z., Mendis, T., Shen, X.: Urban solar utilization potential mapping via deep learning technology: a case study of Wuhan. China. Appli. Energy **250**, 283–291 (2019)
34. Nasrallah, H., et al.: Lebanon solar rooftop potential assessment using buildings segmentation from aerial images. IEEE J. Selected Topics Appli. Earth Observat. Remote Sensing **15**, 4909–4918 (2022)
35. Hossain, M.S., Karlson, M., Neset, T.-S.S.: Application of GIS for cyclone vulnerability analysis of Bangladesh. Earth Sci. Malaysia **3**(1), 25–34 (2019)

Analysis of Energy Efficiency of a Distributed STAR-RIS-Assisted Wireless System

H. Rashmi(✉), Ashvini Chaturvedi, and Bodempudi Naga Siva Prasad

Department of Electronics and Communication Engineering, National Institute of Technology Karnataka, Surathkal 575025, India
{rashmih.217ec012,ashvinichaturvedi,naga.207ec003}@nitk.edu.in

Abstract. This paper analyses the energy efficiency of simultaneously transmitting and reflecting reconfigurable intelligent surfaces (STAR-RIS) assisted wireless systems. The system is analysed for an outdoor scenario specified in the 3GPP Urban Micro (UMi) path-loss model. Two schemes for a distributed STAR-RIS system are studied, namely, the Exhaustive RIS-aided (ERA) scheme and the Opportunistic RIS-aided (ORA) scheme. Performance of three primary operating protocols of STAR-RIS: energy splitting (ES), mode selection (MS) and time switching (TS) protocols are validated in terms of energy efficiency (EE). Channel under study is assumed to follow $\alpha - \kappa - \mu$ fading distribution. The system is analysed for different values of channel fading parameters, namely number of STAR-RIS units and the number of elements in each STAR-RIS. Simulations are carried out in MATLAB, and results are analysed for sub-6 GHz operating frequency.

Keywords: $\alpha - \kappa - \mu$ fading distribution · energy efficiency · distributed STAR-RIS · ERA · ORA · energy splitting · mode selection · time switching

1 Introduction

The present beyond fifth generation (B5G) wireless communication demands higher data rates while complying with ultra-reliable low-latency requirements of certain applications. In sync with these challenging specifications, green communication is one of the major thrust areas of research, considering the increasing carbon footprints due to use of millions of IoT devices in widespread activities, consuming huge amounts of power. Thus, in IoT-based network architecture, energy efficiency (EE) is an important performance indicator as IoT devices' lifetime and network operating cost are of major concerns. The conventional methods, such as massive multiple-input multiple-output (MIMO) and relays, have higher hardware complexity and consume considerable amounts of power. To address the intense power requirement, intelligent reflecting surfaces (IRS) have emerged as a promising technique since IRS are low-power passive elements.

© The Author(s), under exclusive license to Springer Nature Switzerland AG 2026
A. Kannan et al. (Eds.): ADCOM 2025, CCIS 2947, pp. 115–129, 2026.
https://doi.org/10.1007/978-3-032-26269-1_8

IRS enables an energy-efficient smart radio environment by fusing intelligence in network by virtue of steering incident signal with desired phase and amplitude amendments. However, conventional IRS offer limited flexibility due to inherent reflecting-only capability [1].

Simultaneously Transmitting and Reflecting Reconfigurable Intelligent Surfaces (STAR-RIS) provides 360° coverage compared to 180° coverage in conventional reflecting-only RIS. Hence, it enables connectivity to users on both the sides, i.e., reflect side users and transmit side users. There are three key protocols for operating states of STAR-RIS, namely, energy splitting (ES) protocol, mode selection (MS) protocol and time switching (TS) protocol [2]. For non-cooperative distributed STAR-RIS, two schemes are studied for multiple RIS deployment-exhaustive RIS aided (ERA) scheme and opportunistic RIS-aided (ORA) scheme [3].

Reported literature work based on uses of conventional RIS highlights issues such as power optimisation, joint optimisation, spectral efficiency optimisation, and EE optimisation, etc. To attain these tasks, an abstract of various optimisation algorithms is briefed as: In [4], the energy efficiency of the distributed multiple RIS system network is optimised by dynamically turning on and off the RIS, along with optimising the amplitude coefficients. Sequential fractional programming (SFP) approach was adopted in [5] with RIS based wireless system, to provide high energy efficiency and spectral efficiency performance. Authors in [3] maximised EE by cooperatively optimising impedance parameters of RIS panel elements and active beamforming vectors at the transmitter. In [6], outer approximation algorithm was used to solve a nonlinear programming problem to enhance and optimise EE of RIS-assisted network. Authors in [7] optimised EE by combining techniques from alternating optimisation, fractional programming and sequential programming. Gradient descent and trust region-based maximisation algorithms are used to improve energy efficiency in [8], which is proven to be better than SFP algorithm. In [9], secure EE maximisation is proposed using alternating maximisation, sequential fractional programming, and the use of pricing techniques to prevent eavesdropping during signal transmission.

It is observed that EE is analysed and maximised using various stated algorithms, for the RIS-assisted system. However, the impact of STAR-RIS-assisted system on the performance of EE is not analysed. This paper aims to bridge this gap by analysing EE for two schemes of distributed STAR-RIS in a comprehensive manner.

While prior works primarily focus on conventional reflecting-only RIS with EE maximization via optimization algorithms, this work differs in the following fundamental aspects:

1. A mathematical framework is presented for EE analysis of a distributed multi-STAR-RIS-assisted wireless system for a multiple-user scenario.
2. Unlike reflecting-only RIS, STAR-RIS supports simultaneous transmission and reflection, thereby enabling additional protocol dimensions such as ES, MS, and TS, whose comparative EE behaviour remains largely unexplored.

3. Existing EE works focus on optimization-based designs, whereas this paper provides protocol-level comparative EE characterization for distributed multi-STAR-RIS deployments.
4. We analyze EE under $\alpha-\kappa-\mu$ generalized fading, which captures nonlinearity, line-of-sight (LoS) dominance and independant multipath clustering, thus extending beyond Rayleigh/Rician models used in most RIS EE studies.
5. The study explicitly ERA vs ORA selection strategies in distributed STAR-RIS systems from an EE perspective.

Notations: In this paper, scalars are denoted by italics, vectors by boldface lowercase letters, and matrices by boldface uppercase letters. a* denotes the complex conjugate of a. $\mathbb{E}[x]$ denotes expectation of x.

2 System Model

A schematic of a distributed array consisting of multiple STAR RIS for an outdoor scenario is shown in Fig. 1. It comprises a base station (BS), two mobile receivers, UE_1 and UE_2, located at the reflect side and transmit side of STAR-RISs, respectively. End-to-end (E2E) wireless link can be interpreted as a superposition of a LoS component (if prevails) and N number of virtual-LoS (VLoS) components by virtue of multipath signals through N distributed STAR-RIS units. i.e., $R_n, n = 1, 2, ..., N$, where each STAR-RIS consists of L passive STAR elements with L_n as notation for STAR-RIS elements of an arbitrary nth STAR-RIS unit. $\Theta_n = diag([q_{n1}e^{j\theta_{n1}}, ..., q_{nl}e^{j\theta_{nl}}, ...q_{nLn}e^{j\theta_{nLn}}])$. Here, $q_{nl} \in (0, 1]$ represents the amplitude coefficient, $\theta_{nl} \in (0, 2\pi]$ represents the phase-shift of the lth reflecting element of the nth STAR-RIS. Further, it is presumed that the perfect channel state information (CSI) is available and is obtained using channel estimation techniques described in the literature [2].

Let h_{nl}, g_{Rnl} and g_{Tnl} denote the complex channel coefficients for channel segments as; BS to the lth passive element of the nth RIS, STAR-RIS to UE_1 (at reflect side), and STAR-RIS to UE_2 (at transmit side), respectively. It is presumed that a LoS link for entities BS-UE_1 and BS-UE_2 prevails. Let, h_{R0} and h_{T0} denote channel coefficient of the direct LoS between transmitter (BS)-UE_1 and BS-UE_1, respectively. Let x_S denotes a transmitted symbol from BS. Two schemes for multiple STAR-RIS selection, i.e., the Exhaustive RIS-aided (ERA) wireless system and the Opportunistic RIS-aided (ORA) wireless system are briefly discussed next. The various notation used in the manuscript are given in Table 1.

2.1 The Exhaustive RIS-Aided Scheme

In ERA scheme, all the N units of STAR-RIS participates in transmission of symbols between BS and UE_1/ UE_2. The STAR-RIS performs beam-steering to redirect the incident signal to the receivers (UE_1/ UE_2) [3]. The received signal y_R^{ERA} and y_T^{ERA} at UE_1 and UE_2, respectively, is given by

Table 1. Notations and descriptions

Notations	Description
N	Number of STAR-RIS panels
$R_n, n = 1, 2, ..., N,$	Label assigned to each STAR-RIS panel
Θ_n	Phase shift matrix of the nth STAR-RIS
$q_{nl} \in (0, 1]$	Amplitude coefficient of the lth reflecting element of the nth STAR-RIS
$\theta_{nl} \in (0, 2\pi]$	Phase-shift of the lth reflecting element of the nth STAR-RIS
h_{nl}	Complex channel coefficients from the BS to the lth passive element of the nth RIS
g_{Rnl}	Complex channel coefficients from the STAR-RIS to UE_1
g_{Tnl}	Complex channel coefficients from the STAR-RIS to UE_2
h_{R0}	Channel coefficient of the direct link from the BS to UE_1
h_{T0}	Channel coefficient of the direct link from the BS to UE_2
x_S	Symbol transmitted from BS
y_R^{ERA}	Received signal at UE_1 due to ERA scheme
y_T^{ERA}	Received signal at UE_2 due to ERA scheme
P_s	transmit power of the symbol x_S in dBm
w_R	Additive white Gaussian noise at UE_1
w_T	Additive white gaussian noise at UE_2
σ_R^2	Noise variance of the reflect side signal
σ_T^2	Noise variance of the transmit side signal
σ_θ^2	Phase error variance due to quantisation and hardware impairments
$\bar{\rho}_R$	Average SNR at UE_1
$\bar{\rho}_T$	Average SNR at UE_2
y_R^{ORA}	Received signal at UE_1 due to ORA scheme
y_T^{ORA}	Received signal at UE_2 due to ORA scheme
α	Channel fading parameter describing non-linearity of the scattering elements in the environment
κ	Channel fading parameter describing dominance of LOS path
μ	Channel fading parameter describing number of independent clusters of multipath component
$\Gamma(\cdot)$	Gamma function
γ	Average SNR per symbol period
EE	Energy Efficiency
BW	Bandwidth
R_{th}	Target spectral efficiency
P_{tol}	Total power consumed by the STAR-RIS-assisted system
$\tilde{P}_{nl}$	Circuit dissipation power at the lth element of nth STAR-RIS
$\tilde{P}_S$	Circuit dissipation power at the BS
$\tilde{P}_D$	Circuit dissipation power at receiver (UE_1/UE_2)

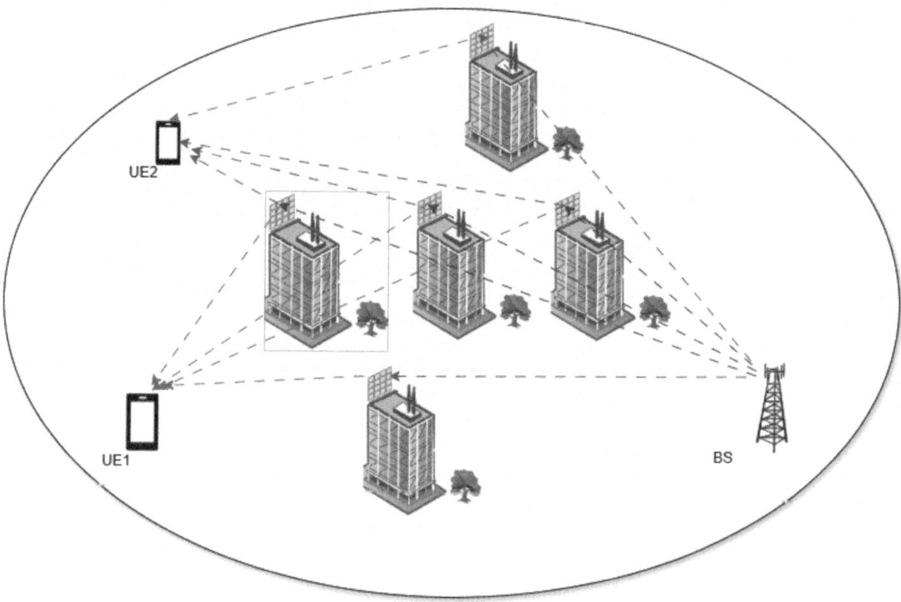

Fig. 1. Illustration of a distributed multiple STAR-RIS-aided wireless system.

$$y_R^{ERA} = \sqrt{P_s}(h_{R0} + \sum_{n=1}^{N}\sum_{l=1}^{L_n} g_{Rnl}q_{nl}e^{j\theta_{nl}}h_{nl})x_s + w_R \tag{1}$$

$$y_T^{ERA} = \sqrt{P_s}(h_{T0} + \sum_{n=1}^{N}\sum_{l=1}^{L_n} g_{Tnl}q_{nl}e^{j\theta_{nl}}h_{nl})x_s + w_T \tag{2}$$

P_s is transmitted power associated with symbol x_S in dBm. w_R and w_T represents the additive white gaussian noise (AWGN) at UE_1 and UE_2, respectively, i.e., $w_R \sim \mathcal{CN}(0, \sigma_R^2)$, $w_T \sim \mathcal{CN}(0, \sigma_T^2)$. Assuming, ideal phase shift at the STAR-RISs, the received signal-to-noise ratio (SNR) at receivers UE_1 and UE_2 is expressed as

$$SNR_R^{ERA} = \bar{\rho}_R|h_{R0} + \sum_{n=1}^{N}\sum_{l=1}^{L_n} g_{Rnl}q_{nl}h_{nl}|^2 \tag{3}$$

$$SNR_T^{ERA} = \bar{\rho}_T|h_{T0} + \sum_{n=1}^{N}\sum_{l=1}^{L_n} g_{Tnl}q_{nl}h_{nl}|^2 \tag{4}$$

where, $\bar{\rho}_R = P_s/\sigma_R^2$, $\bar{\rho}_T = P_s/\sigma_T^2$, denotes the average SNR (in dBm) at UE_1 and UE_2, respectively.

2.2 The Opportunistic RIS-Aided Scheme

In this scheme, instead of engaging all the STAR-RIS units in the transmission, a specific STAR-RIS which offers the highest SNR [3] is considered and participates in E2E transmission of signals. Hence, ORA scheme facilitates a tremendous reduction in resources use and thus turns out to be more energy efficient. Similar to (1) and (2), we can express received signal at users UE_1 and UE_2, as follows

$$y_R^{ORA} = \sqrt{P_s}(h_{R0} + \sum_{l=1}^{L_n} g_{Rnl} q_{nl} e^{j\theta_{nl}} h_{nl}) x_s + w_R \tag{5}$$

$$y_T^{ORA} = \sqrt{P_s}(h_{T0} + \sum_{l=1}^{L_n} g_{Tnl} q_{nl} e^{j\theta_{nl}} h_{nl}) x_s + w_T \tag{6}$$

To comply with the ideal phase shift configuration criterion and optimal SNR criterion, SNR estimate at the receivers UE_1 and UE_2 can be expressed as

$$SNR_R^{ORA} = \max_{1 \le n \le N,} \bar{\rho}_R |h_{R0} + \sum_{l=1}^{L_n} g_{Tnl} q_{nl} h_{nl}|^2 \tag{7}$$

$$SNR_T^{ORA} = \max_{1 \le n \le N,} \bar{\rho}_T |h_{T0} + \sum_{l=1}^{L_n} g_{Tnl} q_{nl} h_{nl}|^2 \tag{8}$$

It must be noted that the number of updates needed for acquisition of precise CSI is the same for the schemes ERA or ORA. However, in the ORA scheme, the receiver (UE_1/UE_2) is required to process fewer signals, i.e., (L_n+1) over $(N \times L_n+1)$ in ERA scheme, hence, yields an improved degree of energy efficiency on deploying ORA scheme.

2.3 STAR-RIS Protocols

The three main protocols of STAR-RIS are ES, MS and TS protocols. In ES protocol, each RIS element splits the incident signal energy simultaneously into a reflect signal and a transmit signal. If $\gamma_r \in [0, 1]$ is the power splitting ratio allocated to the reflecting link, then the power at transmitting end is $\gamma_t = 1 - \gamma_r$. In MS protocol, the RIS elements are divided into two disjoint subsets: one set N_r dedicated to reflection and the other N_t to transmission. In TS protocol, the entire RIS operates in reflection mode for a fraction $\tau_r \in [0, 1]$ of the time and in transmission mode for the remaining fraction $\tau_t = 1 - \tau_r$ [2].

2.4 The $\alpha - \kappa - \mu$ Channel

During simulation study, the wireless channel is treated as an $\alpha - \kappa - \mu$ channel, a generalised fading channel model [10]. The model is suitable for analysing a line-of-sight (LOS) channel. As STAR-RISs provide virtual LOS paths for propagating signal between transmitter and receiver (UE_1/UE_2), the considered channel model is well-suited for the comprehensive analysis. The fading parameters, i.e., α, κ and μ, are defined as follows: $\alpha > 0$ denotes non-linearity of the objects (reflecting and scattering materials) present in environment. $\kappa > 0$ represents the dominance of LOS path, i.e., ratio of total power of the dominant components to total power of the scattered signal component. $\mu > 0$ denotes number of independent clusters of multipath components. Probability density function (pdf) of instantaneous SNR, γ of $\alpha - \kappa - \mu$ channel is given by

$$f_\gamma(\gamma) = \sum_{i=0}^{\infty} \frac{\alpha \mu^{\mu+2i} \kappa^i (1+\kappa)^{\mu+i}}{2\Gamma(\mu+i)\Gamma(i+1)e^{\kappa\mu}\bar{\gamma}^{\frac{\alpha}{2}(\mu+i)}} \gamma^{\frac{\alpha}{2}(\mu+i)-1} e^{-\frac{\mu(1+\kappa)}{\bar{\gamma}^{\frac{\alpha}{2}}}\gamma^{\frac{\alpha}{2}}} \tag{9}$$

In (9), $\Gamma(\cdot)$ is a gamma function and $\bar{\gamma}$ is average SNR per symbol.

2.5 Energy Efficiency

The energy efficiency, in Mb/Joule, of an RIS-assisted system is given [3] by

$$EE = BW \times R_{th}/P_{tol} \tag{10}$$

where BW denotes system bandwidth and R_{th} denotes target spectral efficiency (SE). P_{tol} denotes total power consumed by the STAR-RIS-assisted system and is expressed as

$$P_{tol} = P_S + \sum_{n=1}^{N} \sum_{l=1}^{L_n} \tilde{P}_{nl} + \tilde{P}_S + \tilde{P}_D \tag{11}$$

where $\tilde{P}_{nl}$, $\tilde{P}_S$, $\tilde{P}_D$ denote circuit dissipation power (CDP) components at l^{th} element of nth STAR-RIS, BS and receivers (UE_1 /UE_2), respectively.

Practical Power Consumption Considerations: For practical relevance, indicative hardware power values are considered based on literature [11–13], related to recent RIS hardware prototypes:

- BS transmit power amplifier (PA) efficiency: 35–45 %
- STAR-RIS element controller power: 5–10 mW per element
- STAR-RIS biasing network: 1–3 mW per element
- User-RF front-end + baseband: 100–300 mW

Accordingly, the total power consumption can be expressed as:

$$P_{tol} = \frac{P_S}{\eta_{PA}} + P_{RIS} + \tilde{P}_S + \tilde{P}_D \tag{12}$$

where η_{PA} is PA efficiency, $P_{RIS} = \sum_{n=1}^{N} \sum_{l=1}^{L_n} \tilde{P}_{nl} = N L_n \tilde{P}_{nl}$. For example, if $L = 50$, $N = 5, \tilde{P}_{nl} = 8\ mW$, then $P_{RIS} = 5 \times 50 \times 8mW = 2W$. This will significantly affect P_{tol}, especially for larger values of N and L_n.

Analytical Scaling Insight on Energy Efficiency: In this section, we provide analytical insight into the scaling behaviour of EE with respect to N and L_n.

Under ideal phase alignment, the cascaded channel formed by distributed STAR-RIS units adds coherently. Hence, the effective channel gain can be approximated, [14,15], as

$$\left| \sum_{n=1}^{N} \sum_{l=1}^{L_n} g_{nl} q_{nl} \right|^2 \propto (N L_n)^2. \tag{13}$$

Accordingly, the average SNR scales as

$$\text{SNR} \propto P_S (N L_n)^2, \tag{14}$$

The achievable rate in bits/s/Hz can be expressed using the Shannon capacity formula as

$$R \approx \log_2 \left(1 + c P_S (NL)^2\right), \tag{15}$$

where c is a proportionality constant that depends on channel statistics and noise power. Total power consumption of the STAR-RIS-assisted system can thus be approximated as

$$P_{\text{tol}} \approx P_S + N L_n \tilde{P}_{nl}. \tag{16}$$

Therefore, the energy efficiency scales approximately as

$$\text{EE} \sim \frac{\log_2 \left(1 + c P_S (N L_n)^2\right)}{P_S + N L \tilde{P}_{nl}}. \tag{17}$$

We can infer that, for small values of $N L_n$, the numerator grows faster than the denominator, thereby increasing EE. For large $N L_n$, the linear growth of circuit power dominates, which may result in decrease of EE.

Note on Impact of Imperfect CSI and Phase Errors - The preceding analysis assumes perfect CSI and ideal continuous phase control at the STAR-RIS elements. However, in practical deployments, CSI acquisition errors and discrete phase resolution degrade their performance. If σ_θ^2 denote the variance of phase errors due to quantization and hardware impairments, the effective channel gain scales as $(NL)^2 e^{-\sigma_\theta^2}$ [16,17]. This usually leads to a reduction in the received SNR. However, despite the degradation in absolute EE, the relative performance trends among ERA and ORA schemes remain largely unchanged. This is because ORA uses fewer STAR-RIS units, thereby reducing circuit power consumption. Hence, ORA-based protocols tend to be more energy efficient than ERA-based protocols, even under practical or hardware impairments. A comprehensive robustness analysis incorporating CSI estimation overhead and discrete phase constraints is an interesting research direction for future work.

3 Results and Discussions

In this section, we analyse energy efficiency (EE) for ERA and ORA model of a distributed multiple STAR-RIS system. Analysis is performed for three main protocols of STAR-RIS: ES, MS and TS, while regulating the parameters: number of STAR-RIS panels (N), number of elements in the panel (L), and fading parameters of the channel (α, κ and μ). Simulation parameters are specified in Table 2, unless otherwise specified. Also, non-LoS (nLoS) condition in the 3GPP Urban Micro (UMi) path-loss model is considered.

Table 2. Network elements and signal specifications

System Parameters	Specifications
Operating Frequency [GHz]	5.9
Bandwidth [MHz]	10
Horizontal distance between Source and receivers [m]	100
Number of STAR-RIS (N)	5
Number of elements in each STAR-RIS panel(L)	50
Reflection and transmission coefficients of STAR-RISs	0.7, 0.3
Transmit Power, P_S [dBm]	[0 30]
Antenna Gains of source, receivers and STAR-RISs G_S, G_D, G_{R_n}	[5 5 5]
Target Spectral Efficiency $R_{th}[b/s/Hz]$	1
Thermal noise power density [dBm/Hz]	-174
Noise figure, NF[dBm]	10
α, κ, μ	2, 1, 1

Table 3. Energy efficiency comparison of ERA vs ORA for STAR-RIS protocols

Distributed STAR-RIS Schemes	EE (Mb/J)					
	Achievable rate 45 b/s/Hz			Achievable rate 48 b/s/Hz		
	ES	MS	TS	ES	MS	TS
ERA	63.34	71.04	63.34	48.66	51.75	31.18
ORA	207.42	213.14	177.70	135.03	155.24	74.57

In Fig. 2, ERA and ORA schemes are compared in terms of energy efficiency for three main protocols of STAR-RIS. As observed in Fig. 2 characteristics, ORA scheme outperforms ERA scheme. This is due to the fact that in ERA, all N units of STAR-RIS participate, whereas in ORA, a specific STAR-RIS

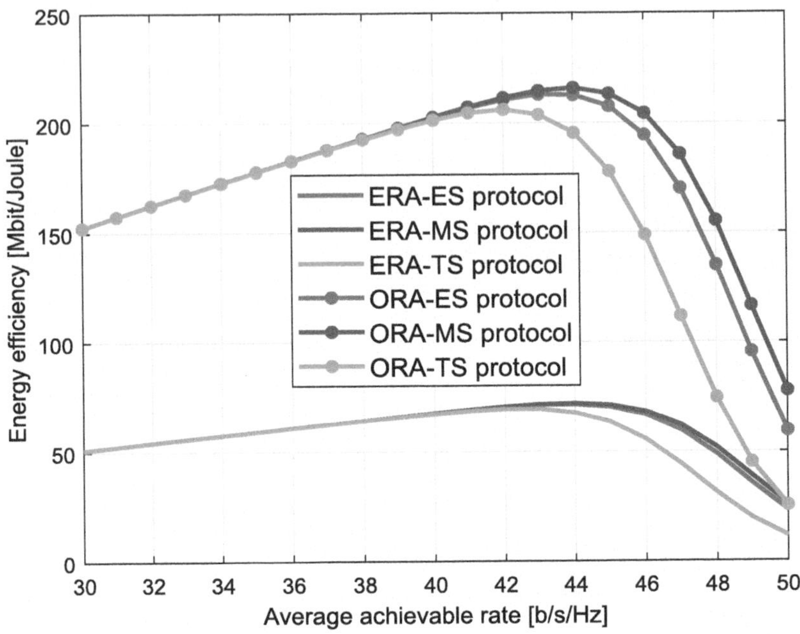

Fig. 2. ERA vs ORA for three major operating protocols of STAR-RIS

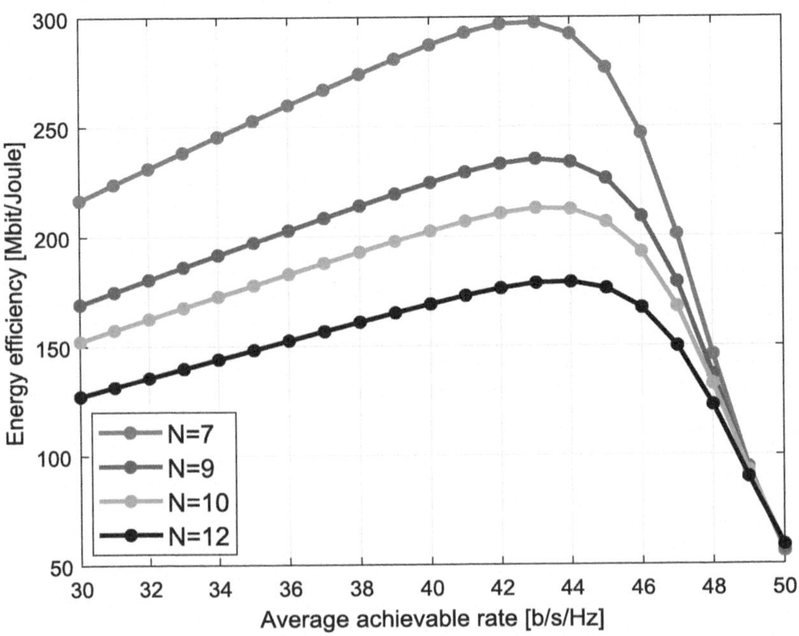

Fig. 3. ORA for varying values of N

that yields the best SNR only gets a chance to participate. It is also observed that among the three operating protocols of STAR-RIS, MS protocol gives better energy efficiency compared to ES and TS. Rationale for this improved EE performance can be explained as: In ES protocol, energy of the incident beam gets split into reflect side signal (towards user UE_1) and transmit side signal (towards user UE_2). This causes reduced power to each signal path, and thus signal strength is lower than the one that resides in beams generated in MS and TS modes. In TS protocol, there is a continuous switching of RIS elements over the stipulated time-duration between transmit and receive sides, resulting in more power consumption while allowing only 50% of the power to each side. However, in MS protocol, there are dedicated elements for reflecting the signal and transmitting the signal. Hence, MS protocol has considerably better energy efficiency. As ORA scheme with MS protocol offers the best energy efficiency, the impact of parametric variations in STAR-RIS features along with a controlled regulation in a channel parameter are analysed for MS protocol with underlying ORA scheme (ORA-MS).

It is also observed that EE behaves differently at a lower achievable rate compared to a higher achievable rate. At low SNR, EE increases with SNR since the achievable rate grows nearly linearly while circuit power remains relatively small. However, at high SNR, the achievable rate increases only logarithmically whereas transmit and circuit power continue to grow linearly. As a result, the improvement in rate becomes marginal compared to the increase in power consumption, leading to diminishing EE at higher SNR. Table 3 summarises Fig. 2, for achievable rates of 45 b/s/Hz and 48 b/s/Hz.

Table 4. Energy efficiency of ORA-MS for different values of N

Number of STAR-RIS panels (N)	EE (Mb/J)	
	Achievable rate 42 b/s/Hz	Achievable rate 46 b/s/Hz
7	296.78	246.83
9	233.08	208.94
10	210.45	192.72
12	176.18	167.33

In Fig. 3, EE performance of ORA-MS is evaluated for different values of N. As N increases, EE reduces. This is evident from (10) and (11). Table 4 summarises Fig. 3, for specific achievable rate 42 b/s/Hz and 46 b/s/Hz. In Fig. 4, EE performance dependency of ORA-MS with respect to variation in number of STAR elements L is analysed. As L increases, EE reduces. The analysis trend is similar to that of varying N. Table 5 summarises Fig. 4, for achievable rates of 42 b/s/Hz and 46 b/s/Hz. Also, the decreasing EE trend with increasing N and L(Figs. 3 and 4) becomes more pronounced under realistic per-element power consumption, reinforcing the need for optimal STAR-RIS sizing. The simulation

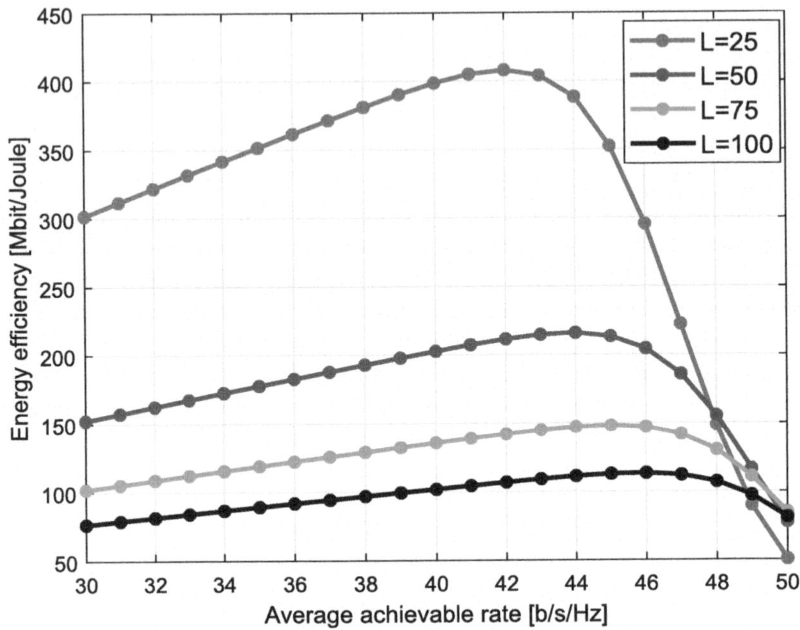

Fig. 4. ORA for varying values of L

Table 5. Energy efficiency of ORA-MS for different values of L

Number of elements in STAR-RIS panel (L)	EE (Mb/J)	
	Achievable rate 42 b/s/Hz	Achievable rate 46 b/s/Hz
25	407.90	295.38
50	211.33	204.28
75	142.06	146.91
100	106.90	113.36

results in Fig. 3 and Fig. 4 are consistent with the derived scaling behavior in Sect. 2.5. In Fig. 5, it is observed that as α increases, EE decreases. The rationale for this specific trend can be illustrated as: with the amount of non-linearities present in the environment increases, scattering increases, resulting in decline in EE. EE analysis for different values of α is summarised in Table 6.

Simulations were also carried out for different values of κ or μ, while keeping other parameters as specified in Table 2. However, there was no impact on energy efficiency. This is because κ or μ have an impact only on the shape of fading and not average power, especially when the power is normalised. To summarise the effect of fading parameters on EE, α controls the fading nonlinearity, resulting in noticeable impact on ergodic rate, resulting in EE variation. However, κ or μ primarily impact only the small-scale fading distribution shape and thus EE becomes insensitive to any variations in κ or μ.

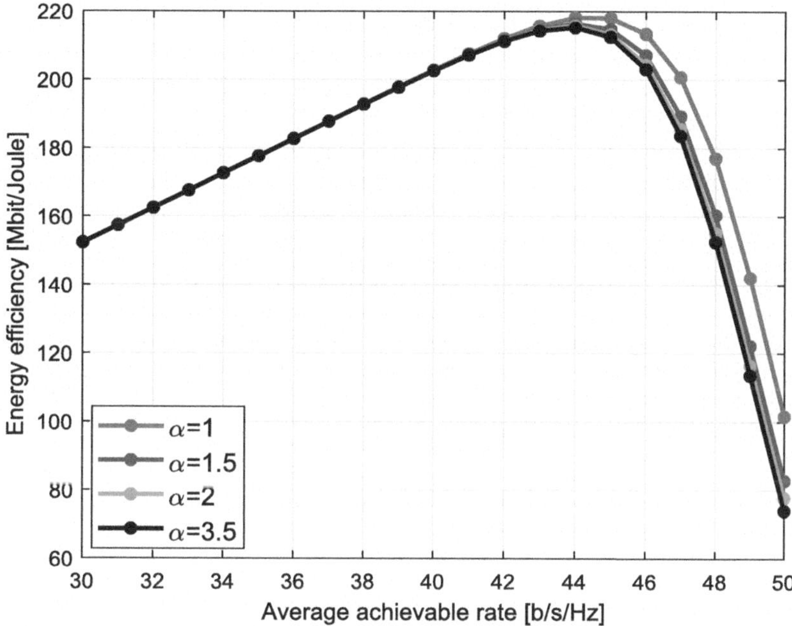

Fig. 5. ORA for different values of α

Table 6. Energy efficiency of ORA-MS for different values of α

Non-linearity measure (α)	EE (Mb/J)	
	Achievable rate 42 b/s/Hz	Achievable rate 46 b/s/Hz
1	213.34	177.04
1.5	207.10.33	160.36
2	204.29	155.24
3.5	203.08	152.63

4 Conclusion

In this paper, two schemes namely, ERA and ORA are analysed especially for energy efficiency (EE) of a distributed wireless system operating with three distinct underlying protocols of STAR-RIS. Unlike prior studies that focus predominantly on reflecting-only RIS and optimization-driven designs, this work provided a protocol-level comparative EE characterization of distributed STAR-RIS systems under a generalized αâĂŞκâĂŞμ fading environment. Based on simulation results, it is observed that ORA offers much better EE than ERA for all three STAR-RIS operating protocols considered. Further, with respect to deploying a particular protocol, the analysis reveals that MS protocol yields better EE compared to TS protocol and ES protocol. Further, results analysis infer that as the number of STAR-RIS panels (N) and number of elements in the panel (L) values increase, the EE drops. With respect to fading parameters, it is keenly observed that varying α has an impact on EE while varying κ or μ has negligible or no impact on EE.

Beyond numerical evaluation, analytical scaling insights were developed to explain the observed EE trends. To improve practical relevance, indicative hardware power consumption factors were discussed and incorporated. Future work involves analysing EE for different generalised fading channels such as $\alpha - \eta - \kappa - \mu$, for the ERA and ORA schemes for various protocols of STAR-RIS. Performance measures such as Ergodic capacity (EC), Symbol error rate (SER) can also be analysed under different channel conditions for the specified schemes and STAR-RIS protocols. Also, we can extend this framework to incorporate detailed hardware-aware models, imperfect channel estimation overhead, and additional generalized fading environments such as $\alpha - \eta - \kappa - \mu$.

Disclosure of Interests. The authors have no competing interests to declare that are relevant to the content of this article.

References

1. Wu, Q., Zhang, S., Zheng, B., You, C., Zhang, R.: Intelligent reflecting surface-aided wireless communications: a tutorial. IEEE Trans. Commun. **69**(5), 3313–3351 (2021). https://doi.org/10.1109/TCOMM.2021.3051897
2. Mu, X., Liu, Y., Guo, L., Lin, J., Schober, R.: Simultaneously transmitting and reflecting (STAR) RIS aided wireless communications. IEEE Trans. Wirel. Commun. **21**(5), 3083–3098 (2022). https://doi.org/10.1109/TWC.2021.3118225
3. Do, T.N., Kaddoum, G., Nguyen, T.L., da Costa, D.B., Haas, Z.J.: Multi-RIS-aided wireless systems: statistical characterization and performance analysis. IEEE Trans. Commun. **69**(12), 8641–8658 (2021). https://doi.org/10.1109/TCOMM.2021.3117599
4. Yang, Z., Chen, M., Saad, W., Xu, W., Shikh-Bahaei, M., Poor, H.V.: Energy-efficient wireless communications with distributed reconfigurable intelligent surfaces. IEEE Trans. Wirel. Commun. **21**(1), 665–679 (2022). https://doi.org/10.1109/TWC.2021.3098632
5. Huang, C., Zappone, A., Alexandropoulos, G.C., Debbah, M., Yuen, C.: Reconfigurable intelligent surfaces for energy efficiency in wireless communication. IEEE Trans. Wirel. Commun. **18**(8), 4157–4170 (2019). https://doi.org/10.1109/TWC.2019.2922609
6. Ma, R., Tang, J., Zhang, X., Wong, K.-K., Chambers, J.A.: Energy-efficiency optimisation for mutual-coupling-aware wireless communication system based on RIS-enhanced SWIPT. IEEE Internet Things J. **10**(22), 19399–19414 (2023). https://doi.org/10.1109/JIOT.2023.3241168
7. Ahsan, M., Jamil, S., Ejaz, M.T., Abbas, M.S.: Energy efficiency maximization in RIS-assisted wireless networks. In: 2021 International Conference on Computing, Electronic and Electrical Engineering (ICE Cube), Quetta, Pakistan, pp. 1–6 (2021). https://doi.org/10.1109/ICECube53880.2021.9628234

8. Rihan, M., Zappone, A., Buzzi, S., et al.: Energy efficiency maximization for active RIS-aided integrated sensing and communication. EURASIP J. Wirel. Commun. Network. **2024**, 20 (2024). https://doi.org/10.1186/s13638-024-02346-8
9. Fotock, R.K., Imoize, A.L., Zappone, A., et al.: Secrecy energy efficiency maximization in dual-metasurface-aided wireless networks. EURASIP J. Adv. Sig. Process. **2025**, 24 (2025). https://doi.org/10.1186/s13634-025-01231-w
10. Prasad Bodempudi, N.S., Chaturvedi, A., Rashmi, H., Kulkarni, M.: Analysis of symbol error probability in GFDM under generalized α-κ -μ channels: an approach based on probability density function for beyond 5G wireless applications. AEU - Int. J. Electron. Commun. **200** (2025). https://doi.org/10.1016/j.aeue.2025.155946
11. https://www.ericsson.com/en/blog/2019/9/energy-consumption-5g-nr
12. Li, Z., Zhang, J., Zhu, J., Dai, L.: RIS energy efficiency optimization with practical power models. In: International Wireless Communications and Mobile Computing (IWCMC). Marrakesh, Morocco 2023, pp. 1172–1177 (2023). https://doi.org/10.1109/IWCMC58020.2023.10183034
13. Wang, J., et al.: Reconfigurable intelligent surface: power consumption modeling and practical measurement validation. IEEE Trans. Commun. **72**(9), 5720–5734 (2024). https://doi.org/10.1109/TCOMM.2024.3382332
14. Zhi, K., Pan, C., Ren, H., Wang, K.: Power scaling law analysis and phase shift optimization of RIS-aided massive MIMO systems with statistical CSI. IEEE Trans. Commun. **70**(5), 3558–3574 (2022). https://doi.org/10.1109/TCOMM.2022.3162580
15. Xu, J., Liu, Y., Mu, X., Schober, R., Poor, H.V.: STAR-RISs: a correlated T&R phase-shift model and practical phase-shift configuration strategies. IEEE J. Sel. Top. Sig. Process. **16**(5), 1097–1111 (2022). https://doi.org/10.1109/JSTSP.2022.3175030
16. Nguyen, N.D., Le, A.-T., Munochiveyi, M., Afghah, F., Pallis, E.: Intelligent reflecting surface aided wireless systems with imperfect hardware. Electronics **11**, 900 (2022). https://doi.org/10.3390/electronics11060900
17. Zhang, Q., Liu, J., Tang, H., Dong, Z., Li, Y.: Practical RIS-aided multiuser communications with imperfect CSI: practical model, amplitude feedback, and beamforming optimization. IEEE Trans. Wirel. Commun. **23**(10), 15245–15260 (2024). https://doi.org/10.1109/TWC.2024.3427695

Multi-task Graph Convolutional Network Framework for Ward-Level Urban Sustainability in Bengaluru

C. A. Prajwal[1] and K. P. Impana[2(✉)]

[1] Department of Computer Science and Engineering, JSS Academy of Technical Education, Bengaluru, India

[2] School of Engineering and Technology, S-Vyasa Deemed to be University, Bengaluru, India

impanaraj@gmail.com

Abstract. Rapid urbanization in Bengaluru has intensified heat stress, air pollution, and unequal access to basic services. This paper presents a prototype framework that integrates ward-level demographic, environmental, and infrastructure features into a *multi-task Graph Convolutional Neural Network with Bayesian uncertainty.* The framework is also combined with explainability techniques and an equity-aware optimization model. It jointly predicts sustainability indicators, identifies key influencing factors such as green space per capita, street lighting, and service equity. It allocates limited greening resources under budgetary and fairness constraints. Application to 18 central wards indicates potential improvements, including a slight reduction in heat intensity ($-0.031\,^{\circ}$C), better air quality (+14.3 points), and enhanced social equity (+0.057). Although the approach is still at an early stage and constrained by small sample size, it offers a proof-of-concept of how uncertainty-aware and interpretable graph learning can support the systematic targeting of urban sustainability measures. The study represents a scalable and transferable model for sustainability planning in rapidly urbanising Indian cities.

Keywords: Graph Convolutional Neural Networks · Urban Sustainability · Multi-task Learning · Bayesian Uncertainty · Explainable AI · Green Space Optimisation · Social Equity · Bengaluru

1 Introduction

Bengaluru, also called the Garden City of India, is today facing many challenges regarding sustainability. The rapid development of the city has led to extreme traffic congestion, more pollution, the loss of lakes and green spaces, and repeated heat stress across many wards. Traditional planning methods cannot solve these very complex issues alone. Innovative platforms like neural networks can be used to predict changes in land use, carbon emissions, and energy demands. This can

© The Author(s), under exclusive license to Springer Nature Switzerland AG 2026
A. Kannan et al. (Eds.): ADCOM 2025, CCIS 2947, pp. 130–143, 2026.
https://doi.org/10.1007/978-3-032-26269-1_9

supply planners with better data on future development [1]. Studies show that machine learning enables energy saving, transport planning, and public health activities within smart cities [2].

Machine learning has also been applied to learn city maps and spatial information. For instance, CNN models can detect latent patterns of cities to inform planning [3]. Another study has revealed that machine learning is assisting in the area of garbage handling, mobility, and effective utilization of energy [4]. Such approaches can influence policies that harmonize development with sustainability for a rapidly growing city like Bengaluru. They also facilitate quick planning. This is achieved by utilizing information coming from multiple sources. Thus, ML can facilitate inclusive development where the advantage reaches more people throughout the city.

Graph Convolutional Neural Networks (GCNNs) provide a class of models that can model the entire city as a graph where the wards can be represented as interrelated nodes. These models can efficiently use both environmental and social data to examine the relations of dependence between various wards. Past studies have shown the effectiveness of GCNs to describe road networks and several urban configurations [5]. This study configures a GCNN especially for Bengaluru by making use of information at the level of the wards, such as population, air quality, and vegetation cover. The model predicts sustainability parameters across the entire city and determines the locations that need priority action [6].

2 Literature Review

Machine learning has been employed in numerous ways to facilitate sustainability. For instance, decision support systems integrate ML with multi-criteria techniques to assist the planner in balancing competing requirements in development [7]. Machine learning models have been deployed to simulate short-term passenger flow within metro systems, assisting transport management [7]. ML is also utilized in energy forecasting of buildings, which facilitates the more efficient use of power by large infrastructures [1]. In ecology, ML has been applied to vegetation mapping and the study of the urban heat island, providing information on environmental changes within cities [8]. Furthermore, ML techniques that utilize various data resources such as POIs and remote sensing can more accurately determine the urban functional zones [9]. These illustrations demonstrate how ML has a broad role to play in making cities sustainable and resilient.

Graph neural networks (GNNs) are ever more central to the study of cities. A recent survey of spatio-temporal GNNs illustrates their use for predicting traffic, environmental science, and public safety [10]. In renewable energy, GCNs have been employed to forecast wind power by utilizing spatial information to improve precision [11]. For planning, GNN-based multi-agent systems offer more justifiable land-use planning by including many stakeholders [12]. Recent urban foundation model studies show that pretrained GNNs can be used for a wide range of sustainability tasks across cities [13]. Finally, hybrid models incorporating temporal attention with GCNs have been shown to improve solutions to climate resilience studies [14]. Individually, these studies show the promise of GNNs as a solid basis for sustainable urban planning with AI.

3 Methodology

This study uses ward-level data from Bengaluru to build a graph-based deep learning model for sustainability planning [15,16]. The dataset contains information on population demography, ward boundaries, public facilities like schools, health centers, playgrounds, and Sanitation facilities, as well as several environmental indicators. From the dataset, the following features were extracted: population density, heat island risk factor, proportion of green cover, air quality indicators, livability indices, and social equity indicators. The wards are modeled as nodes of the graph, whereas adjacency, demographic similarity, and service accessibility relationships join the nodes together to form the edges. The graph is then modeled by a multi-task Graph Convolutional Neural Network (GCNN) with Bayesian uncertainty estimation that simultaneously predicts key sustainability indicators to support the policy development process.

3.1 Data and Feature Engineering

The dataset combines official ward boundaries of BBMP with publicly available service data. From this, we created ward-wise features that reflect sustainability conditions. Population density was calculated by dividing the population of each ward by its land area, while green density was obtained by dividing total green cover by population. Heat island potential was estimated by combining density and green space, giving a measure of thermal stress. Air quality indicators and social vulnerability scores were also derived. These features were normalized before being used as inputs for the graph model. For example, the population density of ward i is defined as:

$$D_i = \frac{P_i}{A_i} \tag{1}$$

where P_{i} is the population and A_{i} is the area of ward i. The heat island potential is expressed as:

$$H_i = \frac{D_i}{\max(D)} - \frac{G_i}{\max(G)} \tag{2}$$

where G_{i} is the green density. This feature engineering ensures that every ward is represented not only by basic demographic data but also by sustainability-related indicators.

Ground Truth Construction and Validation of Sustainability Indicators: Ground-truth sustainability indicators derived from two open dataset authorities for this research were the BBMP Ward-wise Public Goods dataset obtained from OpenCity Portal [15] and the official BBMP ward boundary GeoJSON [16]. The datasets include validated administrative records on public facilities, amenities, lakes, parks, streetlights and other civic assets for each BBMP ward. Public goods attribute data were spatially aligned with ward polygons; each indicator (e.g. green space ratio, service access, vulnerability scores, infrastructure density) was derived based on population and area-normalized formulas. To ensure the derived indicators were trustworthy, all indicators went through rigorous cross-checking to ensure internal consistency, outlier behavior and plausibility via BBMP report patterns and publicly accessible records.

3.2 Graph Construction

Then the city is conceptualized as a graph after the features are generated. The different wards become the nodes, and various relations constitute the edges. Spatial edges join the wards with borders, demographic edges join the wards with the same population density, and social edges join the wards with the same equity/vulnerability status. In general, these various edges enable the graph to reflect both the spatial and the social relations within the entire city. Formally, the graph is given by:

$$G = (V, E, X) \tag{3}$$

where V are the wards, E are the edges, and X are the node features. To model service accessibility, an equity score was introduced that adjusts for distance and fairness, written as:

$$S_i = \frac{1}{d_i} \times E_i \tag{4}$$

where d_{i} is the average distance to services and E_{i} is the equity score for ward i. This representation ensures that Bengaluru's sustainability challenges are encoded in both structural and social terms.

3.3 Multi-task GCNN Model

The GCNN takes the constructed graph as input and learns hidden embeddings for each ward. Using graph convolutional layers, the model updates node features by combining information from neighboring wards. At each layer, the representation of node i is updated as:

$$h_i^{(l+1)} = \sigma\left(\sum_{j \in \mathcal{N}(i)} \frac{1}{c_{ij}} W^{(l)} h_j^{(l)}\right) \tag{5}$$

where $h_{\mathrm{i}}^{(l)}$ is the feature of node i at layer l, $\mathcal{N}(\mathrm{i})$ is its set of neighbors, $c_{\mathrm{i}j}$ is a normalization constant, and $W^{(l)}$ are learnable weights. The model is built to learn different tasks at the same time. It provides predictions on different sustainability indicators like air quality, heat island potential, livability index, service equity, and reduction of vulnerability. To keep up with the imprecise predictions, Bayesian methods along with dropout sampling come into action.

The model therefore provides a prediction as well as how certain it is of making that prediction. Overall training loss is provided as:

$$\mathcal{L} = \sum_{t} w_t \cdot \text{MSE}(y_t, \hat{y}_t) + \lambda \cdot KL \quad (6)$$

where y_t and $\hat{y}_t$ are the true and predicted values for task t, w_t is the weight of that task, and KL is the uncertainty regularization term. This design allows the model to handle multiple sustainability objectives at once and produce reliable outputs.

The proposed architecture uses a Graph Convolutional Neural Network, since ward-level sustainability processes are intrinsically relational in nature rather than purely feature-based. Unlike multilayer perceptrons or convolutional neural networks that assume Euclidean grid structures, GCNNs model spatial adjacency, demographic similarity, and socio-infrastructural connectivity through graph edges that facilitate information propagation across interdependent wards. A multitask formulation is adopted because it assumes that the various indicators of sustainability, such as heat-island risk, air quality, and social equity, are strongly interrelated and share latent explanatory factors. Bayesian dropout allows for uncertainty estimates, which become crucial for small-sample urban analytics, where overfitting is a serious concern. Taken together, these modeling choices result in a principled architecture better positioned to capture the complex, multi-relational, and data-scarce nature of ward-level urban systems.

3.4 Computational Complexity and Scalability

The computational complexity of the multi-task graph neural network primarily depends on the number of edges in the ward-level graph, which is constructed using demographic similarity, spatial adjacency, and k-nearest-neighbour relationships between wards. For each of the eight convolutional layers (GCN, GAT, and TransformerConv), the dominant cost of message passing is $O(|E| \cdot F)$, where $|E|$ is the number of edges and F is the feature dimension, while the node-wise linear projections contribute $O(n \cdot F^2)$ for n wards. With a fixed neighbourhood size $k = 5$, the graph remains sparse and $|E| \approx O(nk)$, so the overall per-layer cost scales approximately linearly in the number of wards.

Because the model is implemented using sparse matrix operations in PyTorch Geometric, the actual computation follows the sparsity pattern of the graph rather than dense $O(n^2)$ operations, which keeps both memory and time overheads manageable even for larger ward populations. When training on multiple ward-level graphs (e.g., many municipal areas), the total training cost grows roughly linearly with the number of graphs, while the cost per ward remains essentially constant. To estimate Bayesian uncertainty via variational inference, the model performs multiple stochastic forward passes; with $m = 5$ Monte Carlo samples, this introduces an approximately $5\times$ multiplicative overhead, but the overall runtime remains practical due to the small, sparse graphs used. Consequently, the proposed multi-task GCNN architecture can scale to larger metropolitan municipalities and to collections of urban graphs while remaining feasible on modest hardware resources.

3.5 Training and Evaluation

The training protocol was consistent across experiments and designed for reproducibility, leveraging deterministic data augmentation to simulate ward-level variability in sustainability indicators. The base dataset of 18 wards was augmented into synthetic samples, yielding approximately 70% for training, 15% for validation, and 15% for testing. Features were scaled using z-score standardization (StandardScaler). The model was trained for up to 200 epochs using the AdamW optimizer (learning rate of 0.0005), with early stopping (patience=50) based on validation loss and complementary learning rate schedulers (ReduceLROnPlateau and CosineAnnealingWarmRestarts). Bayesian uncertainty was estimated via variational inference, with epistemic uncertainty quantified through Monte Carlo sampling (5 forward passes) during evaluation; predictive variance from these samples was computed alongside task-specific performance metrics, including mean squared error (MSE) and the coefficient of determination (R^2). This evaluation framework promotes transparent reporting and facilitates comparisons between the proposed graph neural network and baseline models (Fig. 1).

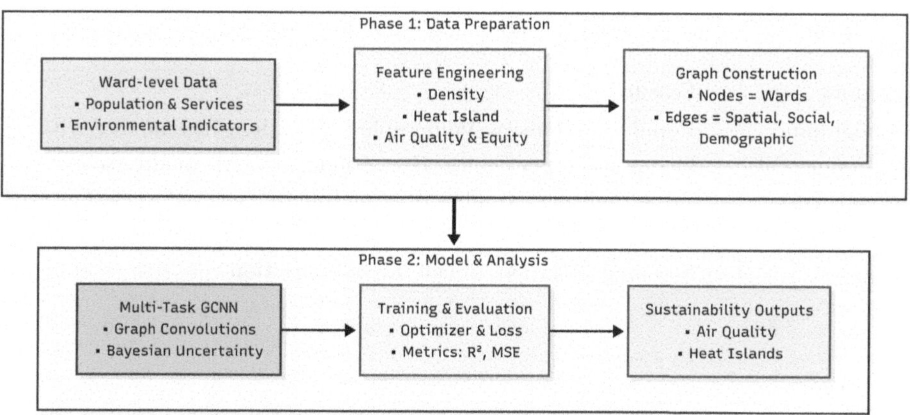

Fig. 1. Flowchart of the proposed methodology.

4 Results

4.1 Model Training and Overall Performance

Training started from a high initial loss and converged to a stable regime. The training log shows initial Train Loss = 1.1315 and Val Loss = 0.7359 (Epoch 1), and the model reached a stable validation loss around 0.395 after early epochs. The Bayesian KL divergence steadily decreased from 7.9445 to about 0.54 by the end of training, indicating reduced posterior uncertainty as training progressed. Given the small study area (18 wards, central Bengaluru) and the high model capacity, the very high R^2 values for some tasks must be read with care; Bayesian uncertainty estimates provide important caution for decision making.

4.2 Feature Extraction and Global Importance

All ward-level features were extracted and normalized before model input. The main feature categories and sample features are listed in Table 1. Permutation-based importance and Spearman correlation measures identify the strongest global drivers of model predictions; the results are in Tables 2 and 3. Street lighting and vulnerable population appear among the top predictors by permutation importance, while green space per capita, street lighting and service equity score show the highest Spearman correlations with model outputs. This indicates that both infrastructure proxies (street lighting) and social indicators (vulnerability, equity) strongly influence the sustainability predictions. The failure of gradient-based importance (due to an RNN backward issue on CPU) required reliance on permutation and correlation methods, which together gave a robust picture of global drivers.

Table 1. Feature set used for modelling (sample categories and features).

Category	Features
Demographic	Population density, Vulnerable population, Income disparity
Environmental	Green space per capita, Heat island risk, Air quality score
Infrastructure	Street lighting, Number of playgrounds, Service accessibility
Social/Equity	Service equity score, Equity entropy, Social vulnerability index
Temporal/Variation	Seasonal variation, Social impact variation

4.3 Local Explanations (LIME) and Feature-Level Detail

Local explanations from LIME give ward-level insight for targeted action. Gradient-based explanations failed in one pass (CPU + cuDNN RNN backward issue), so permutation, Spearman and LIME were used as reliable fallbacks. Table 4 summarises the top LIME contributors for wards 15 (most positive/most negative contributors shown). LIME shows consistent patterns: air quality score and green space metrics often have the largest positive or negative local impact, while playground count and service accessibility frequently appear as local offsetting factors. These ward-level signals are useful for policy because they point

Table 2. Top 10 permutation-based feature importance (higher = more important).

Feature	Importance
Street Lighting	0.049892
Vulnerable Population	0.038434
Green Space Per Capita	0.029820
Population Density	0.028477
Seasonal Variation	0.026848
Equity Entropy	0.025705
Income Disparity	0.023758
Service Accessibility	0.023743
Service Equity Score	0.019977
Number of Playgrounds	0.016896

Table 3. Top 10 Spearman correlation (absolute monotonic relation to outputs).

Feature	Spearman correlation
Green Space Per Capita	0.853457
Street Lighting	0.848735
Service Equity Score	0.830753
Income Disparity	0.752322
Seasonal Variation	0.752322
Population Density	0.752322
Social Impact Variation	0.719545
Social Vulnerability Index	0.714849
Air Quality Score	0.714849
Heat Island Risk	0.714849

to the concrete features to change in each ward (for example, add small parks where green space per capita is low and vulnerable population is high).

Interpretation: air quality and equity-related features drive many local predictions, while infrastructure features (playgrounds, accessibility) act as negative/positive levers depending on local context. LIME highlights where small, low-cost interventions (playgrounds, lighting, service access improvements) can change local sustainability outcomes meaningfully.

4.4 Green Space Optimisation and Policy Implications

The optimisation routine successfully allocated a limited budget to green improvements while prioritising social equity. Key optimisation outcomes are shown in Table 5. Nine wards receive green space allocations (out of 18), with three wards explicitly prioritised for social benefit. The projected system-level

Table 4. Selected LIME local explanations (top contributors) for wards 1–5. Positive values increase the prediction, negative values reduce it.

Ward	Top positive contributor (value)	Top negative contributor (value)
1	Air Quality Score: 0.0382	Number of Playgrounds: −0.0278
2	Air Quality Score: 0.0401	Number of Playgrounds: −0.0287
3	Equity Entropy: 0.0341	Air Quality Score: −0.0434
4	Air Quality Score: 0.0357	Service Accessibility: −0.0338
5	Air Quality Score: 0.0386	Number of Playgrounds: −0.0279

improvements include a 0.031°C average reduction in heat island effect from the allocated greening and an estimated 14.3-point improvement in air quality score for affected wards. Green allocation is expected to raise the overall sustainability index by 0.113 and improve equity by 0.057. Gini for green distribution is 0.553 indicating moderate inequality in green space distribution before optimisation; post-allocation equity measures show measurable improvement.

Table 5. Green space optimisation results and social impact (summary).

Metric	Value
Optimization success	True
Total budget used (units)	50.0
Wards receiving green space	9
Maximum ward allocation (units)	9.7
Projected heat island reduction (°C)	0.031
Projected air quality improvement (points)	14.3
Projected equity improvement	0.057
Projected vulnerability reduction	0.041
Overall sustainability improvement	0.113
Social wards prioritised	3
Green space Gini (distribution)	0.553
Green social benefit (index)	0.833

Budget-Constrained Equity-Aware Optimization: The optimization component is formulated as a nonlinear multi-objective resource allocation problem solved via sequential least squares programming (SLSQP). For each ward i, the decision variable g_i represents green infrastructure allocation, subject to the total budget constraint $\sum_i g_i \leq B$ and non-negativity bounds $g_i \geq 0$.

Equity is operationalized through a social constraint prioritizing wards with high vulnerability ($socialvulnerabilityindex$) and low service equity, alongside an objective that balances predicted heat island reduction, air quality improvement, social equity gains, and vulnerability reduction with tunable *socialweight*. The model first predicts sustainability metrics and optimal allocation via its *optimizationhead*, which initializes the optimization; SLSQP then refines this allocation to maximize the combined multi-objective while satisfying constraints. This approach ensures equitable green space distribution that explicitly favors underserved wards.

Policy Insight: optimising green allocation with social constraints yields measurable benefits for heat reduction, air quality and social equity. Given the feature importance results, policies that combine greening with improvements to street lighting, service accessibility and small-scale playgrounds will have a larger effect in wards with high vulnerability and low green space per capita. Figure 2 visualizes the spatial distribution of heat island intensity across wards, highlighting concentrated hotspots in high-density areas. Figure 3 illustrates the uneven distribution of social equity across wards, motivating the need for equity-aware optimization. Figure 4 shows the optimized allocation of green infrastructure, with higher allocations directed toward vulnerable and undeserved wards.

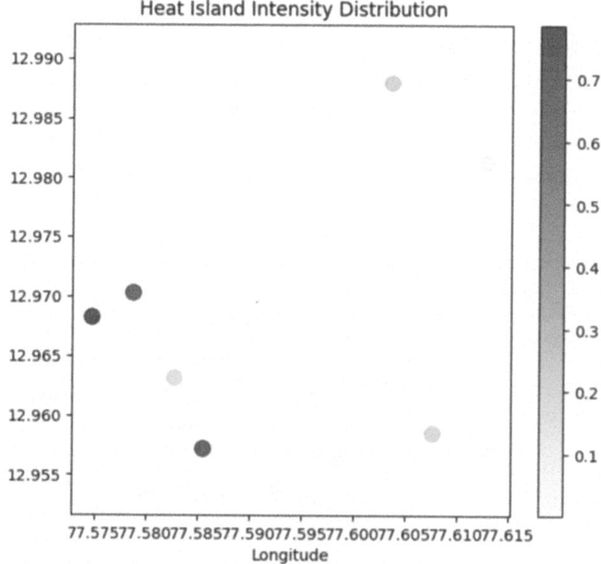

Fig. 2. Ward-level heat island intensity distribution across central Bengaluru.

4.5 Baseline Comparison with Traditional Methods

To contextualise the performance of the multi-task GCNN, the model was compared against a set of traditional and non-graph baselines trained on the same

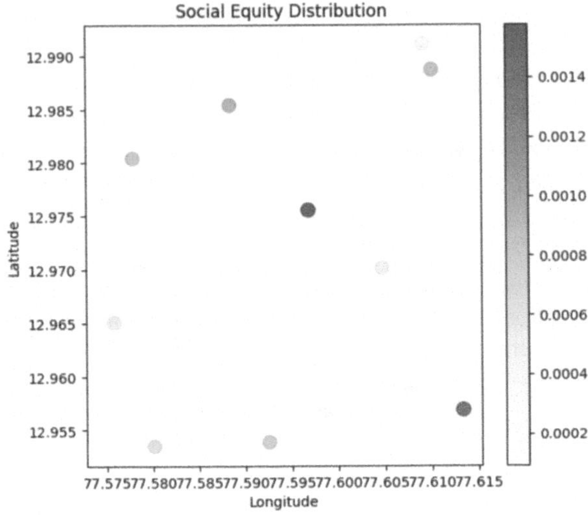

Fig. 3. Ward-level social equity distribution across central Bengaluru.

ward-level features and target indicators. The baselines include ordinary least squares (OLS) linear regression, Ridge and Lasso regression, a Random Forest regressor, a LightGBM model, and a multilayer perceptron (MLP). All models were evaluated on the same held-out test wards using the coefficient of determination (R^2) and mean absolute error (MAE) for four tasks: parks, heat island, air quality, and social equity (Tables 6 and 7).

Table 6. Baseline R^2 comparison across sustainability prediction tasks.

Model	Parks	Heat Island	Air Quality	Social Equity
OLS	0.8713	0.0000	0.0000	0.9961
Ridge	0.8503	0.0000	0.0000	0.9332
Lasso	0.9173	0.0000	0.0000	−1.7737
Random Forest	0.3816	0.0000	0.0000	0.2731
LightGBM	−1.0493	0.0000	0.0000	−1.7737
MLP	−0.8376	0.0000	0.0000	–
Proposed GCNN	**0.9980**	**1.0000**	**1.0000**	**0.9980**

Across tasks, the GCNN achieves consistently higher R^2 and lower MAE than the non-graph baselines, especially for parks, heat island, and air quality. In this setting, an R^2 value of 0 indicates that the model is no better than predicting the mean of the training data on the tiny test split; it is therefore *worse* than the GCNN scores close to 1.0, not better. Negative R^2 values (e.g., Lasso and

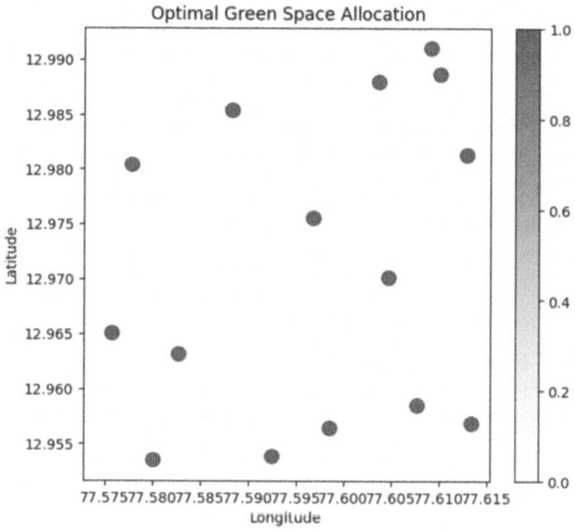

Fig. 4. Optimal green space allocation under budget and equity constraints. (Color figure online)

Table 7. Baseline MAE comparison across sustainability prediction tasks.

Model	Parks	Heat Island	Air Quality	Social Equity
OLS	9.2459	0.0227	1.1355	0.0000
Ridge	10.1647	0.0174	0.8719	0.0001
Lasso	8.0154	0.0673	0.0673	0.0005
Random Forest	17.2467	0.0067	0.5172	0.0003
LightGBM	33.8333	0.1826	9.1282	0.0005
MLP	34.4980	0.1096	18.8715	0.2246
Proposed GCNN	**0.8080**	**0.0020**	**0.1540**	**0.0000**

LightGBM for social equity, some MLP outputs) mean that the model performs worse than this mean baseline and reflect severe overfitting or numerical instability given the extremely small test set. The near-zero MAE values for social equity arise because this indicator is defined on a very narrow scale; differences between models are more evident in their R^2 behaviour and in the other tasks. Overall, these baselines show that standard tabular and spatial models struggle to match the GCNN on multi-task ward-level sustainability prediction, while also underlining the risk of overfitting in such a small-sample regime.

4.6 Limitations and Practical Guidance

The proposed model achieved high prediction accuracy with low uncertainty for most urban sustainability indicators, including Parks ($R^2 = 0.998$, MAE $= 0.808$,

Uncertainty = 0.082), Heat Island (R^2 = 1.000, MAE = 0.002, Uncertainty = 0.068), Air Quality (R^2 = 1.000, MAE = 0.154, Uncertainty = 0.268), Social Equity (R^2 = 0.998, MAE = 0.000, Uncertainty = 0.113), and Vulnerability Reduction (R^2 = 1.000, MAE = 0.001, Uncertainty = 0.267). The improvements in Social Equity notably enhance access for vulnerable groups, fostering community cohesion and reducing inequality.

Model performance metrics (very high R^2 on several tasks) likely reflect the small sample size (18 wards) and the rich feature set; such near-perfect performance can indicate overfitting. The Bayesian uncertainty outputs (KL divergence and task uncertainties) should be used alongside point estimates when advising policy. For practical deployment in Bengaluru, the model outputs are best used as a decision support layer: the optimization recommendations and LIME explanations provide actionable starting points for field surveys and pilot greening interventions, after which the model can be re-calibrated with new data.

Although this framework has excellent potential with respect to prediction and optimization, it also has some important limitations that need to be identified. The framework suffers from a high potential for overfitting due to its very small sample size (18 wards) and therefore provides an overly optimistic assessment of the model's predictive capabilities. Because of this, the predictions made by the model should be evaluated as insights rather than concrete conclusions. Future research should increase the scope of the current study to improve the generalization of findings by considering factors such as temporal variations, remote-sensing integration, and cross-city evaluations. Therefore, while the work currently represents a prototype decision support system, all of the recommendations made by this prototype need to be validated using field surveys, longitudinal data collection, and retraining of the model using larger and more diverse datasets.

5 Conclusion

The report introduces a prototype framework that integrates multi-task Bayesian graph learning with explainable analysis and equity-based optimization for ward-level sustainability assessment in Bengaluru and finds that interventions on green space, street lighting, and service accessibility could provide measurable gains on urban resilience and equity.

The findings here should be treated as preliminary or suggestive rather than conclusive, given the small dataset and stage of application. Future research can go beyond the analysis of the wards, include consideration for temporal connectivity, and compare the analysis with other baselines. This would further substantiate the concrete utility and reliability of the outcomes of the framework, and deepen the possibilities of becoming a feasible decision-support tool for sustainable urban governance.

References

1. Yu, M., Xu, F., Hu, W., Sun, J., Cervone, G.: Using long short-term memory (LSTM) and internet of things (IoT) for localized surface temperature forecasting in an urban environment. IEEE Access **9**, 137406–137418 (2021)
2. Zhou, Z., Wu, X., Peng, B.: An approach to predicting urban carbon stock using a self-attention convolutional long short-term memory network model: a case study in Wuhan urban circle. Remote Sensing **16**(23), 4372 (2024). https://doi.org/10.3390/rs16234372
3. Mansouri, A., Erfani, A.: Machine learning prediction of urban heat island severity in the midwestern United States. Sustainability **17**(13), 6193 (2025). https://doi.org/10.3390/su17136193
4. Kong, G., Peng, J., Corcoran, J.: Modelling urban heat island effects: a global analysis of 216 cities using machine learning techniques. Comput. Urban Sci. **5**, 18 (2025). https://doi.org/10.1007/s43762-025-00178-w
5. Ma, D., He, F., Yue, Y., Guo, R., Zhao, T., Wang, M.: Graph convolutional networks for street network analysis with a case study of urban polycentricity in Chinese cities. Int. J. Geogr. Inf. Sci. **38**(5), 931–955 (2024). https://doi.org/10.1080/13658816.2024.2321229
6. Wang, H., Tang, J., Zhang, J., Yang, J.: Intra-city scale graph neural networks enhance short-term air temperature forecasting, EGUsphere [preprint] (2025). https://doi.org/10.5194/egusphere-2025-3429
7. Wang, Z., Ren, F.: Developing a decision support system for sustainable urban planning using machine learning-based scenario modeling. Sci. Rep. **15**, 13210 (2025). https://doi.org/10.1038/s41598-025-90057-5
8. Delgado-Enales, I., Lizundia-Loyola, J., Molina-Costa, P., Del Ser, J.: A machine learning approach for the efficient estimation of ground-level air temperature in urban areas. Urban Climate **61**, 102415 (2025)
9. Chen, Y., et al.: Semantic-enhanced graph convolutional neural networks for multi-scale urban functional-feature identification based on human mobility. ISPRS Int. J. Geo Inf. **13**(1), 27 (2024). https://doi.org/10.3390/ijgi13010027
10. Jin, G., et al.: Spatio-temporal graph neural networks for predictive learning in urban computing: a survey. arXiv:2303.14483 (2023)
11. Liu, Z., Ware, T.: Capturing spatial influence in wind prediction with a graph convolutional neural network. Front. Environ. Sci. **10** (2022). https://doi.org/10.3389/fenvs.2022.836050
12. Zou, J., Wang, L., Yang, S., Lacasse, M., Wang, L.: Predicting long-term urban overheating and their mitigations from nature-based solutions using machine learning and field measurements. arXiv:2502.18647 (2025)
13. Zhang, W., et al.: Towards urban general intelligence: a review and outlook of urban foundation models. arXiv:2402.01749 (2025)
14. Yu, Y., Li, P., Huang, D., Sharma, A.: Street-level temperature estimation using graph neural networks: performance, feature embedding and interpretability. Urban Climate **56**, 102003 (2024). https://doi.org/10.1016/j.uclim.2024.102003
15. DataMeet Community: Spatial data of municipalities (Maps)—Bangalore: BBMP ward boundaries (GeoJSON). Municipal spatial data project (2016). https://github.com/datameet/Municipal_Spatial_Data. Licensed under CC BY 4.0
16. OpenCity and DataMeet: BBMP ward-wise public goods data. OpenCity data portal (2018). https://data.opencity.in/dataset/bbmp-ward-wise-public-goods-data. Public Domain

Machine Learning and Natural Language Processing

TamilBookAnnot: A Strategic e-Book Annotation Framework in Tamil Integrating Semantic Intelligence Interfaced with Knowledge

S. A. Mohammed Salman[1], Gerard Deepak[2(✉)], and A. Santhanavijayan[3]

[1] Department of Metallurgical and Materials Engineering, National Institute of Technology, Tiruchirappalli, India

[2] Department of Computer Science and Engineering (Cyber Security), Dayananda Sagar Academy of Technology and Management, Bangalore, India
gerard.deepak.christuni@gmail.com

[3] Department of Computer Science and Engineering, National Institute of Technology, Tiruchirappalli, India

Abstract. There is a strategic need for e-book annotation frameworks in regional languages such as Tamil in the context of the Web 3.0 ecosystem, where semantic intelligence, knowledge structuring, and metadata-driven processing form the backbone of digital content organization. This paper presents Tamil Book Annot, a strategic e-book annotation framework in Tamil that integrates semantic artificial intelligence with large language models to enrich annotations through incremental auxiliary knowledge aggregation. Informative dataset terms are extracted and employed for metadata generation, which is subsequently classified using BERT to yield structured semantic entities. Linguistic corpora analysis is performed using Tamil WordNet, while generative enrichment is facilitated through the Tamil large language model (LLM) Gemma 7B to produce auxiliary textual representations. Quantitative semantic reasoning is achieved through similarity measures including Sim Rank and Lloyd's Index, and solution refinement is carried out using a concept-driven metaheuristic optimization algorithm, namely Charged System Search (CSS). The framework integrates semantic intelligence, generative artificial intelligence, and optimization-driven reasoning to generate linguistically coherent and semantically enriched Tamil annotations. Experimental evaluation on a curated corpus of 110 Tamil e-books demonstrates strong annotation quality, achieving 96.89% precision, 97.98% recall, and 97.44% accuracy, confirming the effectiveness of the proposed framework for Tamil e-book annotation in Web 3.0 environments.

Keywords: Generative AI · Knowledge Graph Integration · Tamil e-book Annotation · Tamil Computing

© The Author(s), under exclusive license to Springer Nature Switzerland AG 2026
A. Kannan et al. (Eds.): ADCOM 2025, CCIS 2947, pp. 147–159, 2026.
https://doi.org/10.1007/978-3-032-26269-1_10

1 Introduction

The Web 3.0 has created a demand for frameworks that can support semantic indexing, annotation, and recommendation of digital content for regional languages like Tamil. While book recommendation frameworks have gained attention in the recent past, e-book annotation frameworks remain scarce, particularly in regional languages. Existing frameworks are data-driven but lack knowledge-centric schemes that integrate linguistic, semantic, and auxiliary knowledge. The absence of these factors becomes a limitation when considering the long-term need for structural metadata and machine-interpretable annotations in Web 3.0. Tamil, one of the world's oldest living languages with a strong global presence, continues to have very limited representation in such annotation frameworks. The need to develop Tamil e-book annotation schemes is twofold: first, to ensure that Tamil literary and scholarly works are systematically indexed, preserved, and retrievable, and second, to contribute Tamil linguistic entities as auxiliary knowledge within Web 3.0. By achieving this, Tamil can actively enrich the semantic web envisioned by Tim Berners-Lee, where even regional languages form part of the metadata. Thus, a strategic e-book annotation framework in Tamil is essential for knowledge organization in the era of Web 3.0.

Motivation: The primary motivation of the proposed framework is lack of e-book annotation frameworks in the Web 3.0, specifically for regional languages like Tamil. Semantic intelligence with generative AI aggregation is also the need of the hour, which is along with a strong deep learning model. Metadata-driven and knowledge-centric schemes form the backbone of Web 3.0, yet their integration into regional languages is extremely rare. Hence, the design of the proposed framework addresses this gap by embedding semantic intelligence, generative enrichment, and knowledge-based strategies into a strategic Tamil e-book annotation framework.

Contribution: The primary contributions of the proposed framework is the generation and enrichment of auxiliary knowledge from the dataset perspective in the Tamil language. This is achieved through linguistic-corpora analysis using Tamil WordNet and the generation of summarized text through the Tamil LLM built on Gemma 7B. Another contribution is the incorporation of English metadata by translating Tamil dataset terms into English and integrating them into a community-verified knowledge pipeline, thereby enhancing the auxiliary knowledge density of the dataset terms and making them more informative. The use of Lloyd's Index and Charged System Search (CSS) algorithm provides a metaheuristic-driven optimization mechanism within the framework. Combined with metadata generation and classification using BERT, as well as SimRank and Lloyd's Index with empirically set differential thresholds and step-deviance measures, the framework integrates semantics into a strong quantitative reasoning. This amalgamation of linguistic analysis, auxiliary knowledge augmentation, semantic reasoning, and optimization strategies makes the proposed framework robust.

Organization: The paper is structured as follows. Section 2 provides an overview of Related Works, Sect. 3 discusses the Proposed Methodology, Sect. 4 presents the Performance Evaluation and Results, and Sect. 5 concludes with insights.

2 Related Works

Lacic et al. [1] worked on tag recommendation methods for labeling e-books, introducing a new way to measure semantic similarity. They combined user search terms from Amazon with tags assigned by editors to help tackle the vocabulary gap between readers and publishers. By testing various recommendation algorithms, they found that using hybrid methods, which blend these vocabularies, led to better tag recommendations in terms of accuracy, diversity, and relevance. Hwang et al. [2] created Q-Book, a multimedia annotation tool designed for e-books. This tool lets users enhance digital content with different types of media annotations. It also features interactive capabilities, allowing multimedia tagging that boosts reader engagement and helps with knowledge retention. Their framework shows how adding semantic multimedia annotations can turn static e-books into lively, richly annotated resources for various learning settings. Ma et al. [3] came up with a framework that uses large LLMs to create concept maps from e-books, addressing challenges in automated knowledge extraction. By employing LLMs to pinpoint and organize key concepts and their relationships, this method allows for a dynamic and semantic representation of content in e-books. Chang et al. [4] team introduced a deep learning co-training framework for classifying e-books, which combines several neural networks to learn different features from book metadata and content. Their method outperforms traditional classifiers by effectively using deep representations, which boosts categorization accuracy. This work highlights how a deep understanding through AI can improve the organization and retrieval of large collections of digital books. Indranandita et al. [5] developed a journal classification and search system that merges the Naive Bayes algorithm with a vector space model. This system helps in efficiently categorizing and retrieving documents. Their hybrid approach balances probabilistic classification with vector-based similarity, making it capable of handling a wide range of academic content. This method is a practical blend of classical machine learning and semantic search for managing scholarly documents. Brouard et al. [6] explored a less conventional method for classifying documents using a simple neural network that computes an 'echo' to capture document features. Even with its simplicity, this approach performs well in terms of classification, showcasing alternative methods for applying neural models to semantic text categorization with minimal overhead. Zhang et al. [7] conducted a comparative study assessing TF-IDF, Latent Semantic Indexing (LSI), and multi-word extraction techniques within text classification scenarios. Their findings provide useful insights into the use cases for different feature extraction methods, guiding how to design effective semantic classifiers that balance keyword significance with underlying semantic structures. Devika et al. [8] reviewed different approaches to book recommendation systems. They highlighted the impact of semantic technologies in enhancing recommendations by capturing richer contextual and domain knowledge. Their survey shows the benefits of incorporating semantic reasoning into recommendation frameworks to provide more relevant and personalized book suggestions.

3 Proposed Methodology

Figure 1 illustrates the system architecture of the proposed e-book annotation framework in Tamil, which integrates semantic intelligence with interfaced knowledge. The framework primarily focuses on annotation of Tamil e-books, forming the core dataset comprising various Tamil books. From this dataset, terms and categories are extracted directly from the text using iNLTK, where keyword-based judgment facilitates identification. These extracted terms and categories are then subjected to linguistic knowledge integration by feeding them into the Tamil WordNet, which acts as a lexico-syntactic repository. The contextual linguistic analysis of these terms, along with their hierarchical document structures derived from the Tamil WordNet, is then passed into a Gemma 7B, specifically the Gemma 7B. Gemma 7B is a lightweight, powerful open-weight LLM, optimized for better efficiency, and adaptability across different Indian languages. Its design makes it easy to fine-tune on various datasets. When it comes to Tamil, Gemma 7B has shown some positive results after being fine-tuned with carefully selected multilingual datasets. Community initiatives like Navarasa and its follow-up, Navarasa 2.0, have tailored Gemma 7B for Tamil by instruction-tuning on both translated and native text collections. Consequently, this model has shown significant improvements in comprehending and generating Tamil text, handling tasks such as question answering, summarization, etc. This places Gemma 7B as a strong model for developing high-quality Tamil applications in areas like digital content creation. This plays a crucial role in generating auxiliary knowledge by producing large-scale summary texts. The generated summary text is further parsed in Tamil using iNLTK to yield additional terms, including captions and terminologies that strongly align with the dataset. Subsequently, the Tamil e-book dataset is processed again to extract terms, particularly labels available in both Tamil and English. These terms are translated to English using the Lingvanex Translator API, enabling the generation of knowledge graphs via Google's Knowledge Graph API. The derived entities are then funneled into the CYC pipeline, a commonsense web repository, to enrich auxiliary knowledge, which is further enhanced by population into the YAGO model with additional instances and auxiliary knowledge harvested from the web. This process is performed in English, as auxiliary knowledge density is considerably higher in English than Tamil, and the results are later translated back into Tamil through the Lingvanex API. In parallel, the dataset undergoes TF-IDF analysis in Tamil to identify frequent and informative terms within the document corpus. TF-IDF is a statistical measure used to evaluate the importance of a word in a document relative to a collection of corpus. TF-IDF balances common and rare words to highlight the most meaningful terms. Term Frequency (TF),

Equation 1, measures how frequently a word appears in a document. A higher frequency suggests greater importance that the term is more likely relevant to the document's content. It is given by the formula:

$$\mathrm{TF(t, d)} = \frac{\text{Number of times term t appears in document d}}{\text{Total number of terms in document d}} \tag{1}$$

Inverse Document Frequency (IDF), reduces the weight of common words across multiple documents while increasing the weight of rare words. If a term appears in fewer documents, it is more likely to be meaningful and specific to the context. It is given by the formula:

$$\mathrm{IDF(t, D)} = \log\left(\frac{\text{Total number of documents in corpus D}}{\text{Number of documents containing term t}}\right) \quad (2)$$

The logarithm is used to reduce the effect of very large or very small values, ensuring that the IDF, Eq. 2, scores are scaled appropriately. It also helps balance the impact of terms that appear in very few or extremely many documents. These terms are translated into English using Lingvanex and used to generate structured metadata from Web 3.0 resources. Since Web 3.0 primarily comprises linked open data in English, translation becomes essential for large-scale metadata generation. Given the large, heterogeneous, and high-density nature of the metadata, it is subjected to classification to make it more atomic and manageable. BERT (Bidirectional Encoder Representations from Transformers) is employed as the classifier of choice, and from the classified metadata instances, only the top 40% are selected to preserve domain relevance while avoiding noise. BERT is a language model that understands the context of words by considering them in relation to each other from both sides (bidirectional). This approach allows it to grasp the meanings of whole sentences, not just individual words. So, when we use BERT for classification, we're essentially training it to look at a piece of text and decide which "label" or "category" it fits into, like determining if a sentiment is positive or negative, or if an email is spam or not.

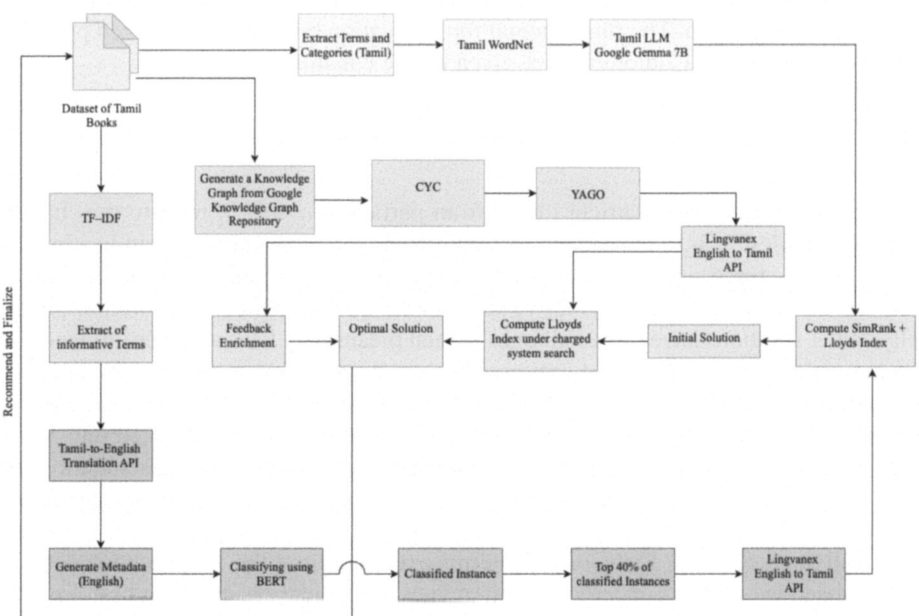

Fig. 1. Proposed TamilBookAnnot e-book annotation framework

To make this work, BERT is typically fine-tuned on a labeled dataset. At the start of every input sentence, we add a special token called CLS. After processing the input, BERT generates a vector for this CLS token that reflects the overall meaning of the text. This vector then goes through a little neural network layer, which is the classifier that gives us the final prediction. These selected instances are translated back into Tamil using Lingvanex and aligned with entities generated by the Tamil LLM (Gemma 7B). To establish semantic consistency, SimRank and Lloyd's Index are computed between these sets of entities. SimRank is fixed at a threshold of 0.80, while Lloyd's Index is empirically set with a step-deviance measure of 0.10. This stringent configuration ensures that only highly cohesive and contextually meaningful entities are retained. However, given that the initial solution set is still significantly large, a metaheuristic optimization algorithm, namely the Charged System Search (CSS), is employed. The CSS, Eq. 3, algorithm recursively explores all entities in the solution space, applying Lloyd's Index as the criteria function with the 0.10 step-deviance to refine the solution until the optimal set is achieved. The CSS algorithm is a nature-inspired optimization algorithm developed by Kaveh and Talatahari in 2010. This method is based on the fact of how charged particles behave in an electric field. In nature, these particles exert forces on each other based on Coulomb's law, which states, like charges repel each other, while opposite charges attract each other. CSS takes this concept and uses it to steer a group of particles (which are basically initial solutions) toward the best solution for an optimization challenge. Particles stand for potential solutions, charge reflects their quality (fitness), force is the interaction that drives both exploration and exploitation, and position is the solution vector being fine-tuned. By repeating this process, the system gradually improves for a nearly optimal solution. The fundamental force equation between two charged particles in CSS is grounded in Coulomb's law, given by the equation:

$$\mathrm{F_{ij}} = \mathrm{k} \cdot \frac{\mathrm{q_i} * \mathrm{q_j}}{\mathrm{r_{ij}^2}} \cdot \hat{r}_{\mathrm{ij}} \tag{3}$$

Here, F_{ij} is the force that particle i feels from particle j, q_i and q_j refer to the charges of those particles, r_{ij} is the distance separating them, $\hat{r}_{ij}$ is the unit vector indicating the direction of the force, and k is a constant to keep things proportional. In CSS, the charges aren't actual physical quantities, they're based on the quality of the initial solutions. Basically, better solutions have higher charges, which means they have a bigger influence on the annotation process. The final optimized solution set is further enriched by integrating auxiliary knowledge from Google Knowledge Graph API, the CYC–YAGO pipeline, and other external sources. The enriched English entities are then translated into Tamil using Lingvanex and incorporated into the optimal solution set through semantic similarity computation. Once semantic similarity has been established, the enriched and optimized entities are finalized as Tamil annotations. These annotations are then systematically mapped back to the original terms extracted from the Tamil e-book dataset. The finalized annotations undergo domain review for validation, ensuring both linguistic fidelity and semantic precision. In this manner, the framework ensures that e-book annotations in Tamil are enriched with semantic intelligence, interfaced with auxiliary knowledge, and optimized for accuracy and contextual relevance. The TamilBookAnnot framework was evaluated using a unified large-scale Tamil e-book annotation dataset curated from multiple publicly available digital repositories. The primary dataset component was derived from the Tamil Books NA collection (J, 2005) hosted on HuggingFace, which contains digitally formatted Tamil literary works spanning diverse genres. To enrich this base collection, an additional set of approximately 20 books was collected from tamilbookspdf.com, including notable titles such as Kaaval Kottam, Vishnupuram (B. Jeyamohan), Mohini Theevu (Kalki), Sila Nerangalil Sila Manithargal (Jayakanthan), En Iniya Iyanthira (Sujatha), Thuppariyum Sambu (Devan), Kural (Thiruvalluvar), Kolaiyuthir Kaalam (Sujatha), Solaimalai Ilavarasi (Kalki), Ponniyin Selvan, Alai Osai, Cilappatikaram, Manimekalai, Ramavataram (Kambar), Thiruvasagam, and Periya Puranam. To further improve diversity and coverage, a curated subset of titles was also sourced from the Goodreads list "Best Tamil Classics You Must Read," capturing both modern and classical Tamil literature.

All books were processed in a uniform e-document format, cleaned for font variations, and structural variations. The initial annotation layer was derived by extracting the top 5% of glossary-frequency terms from each document, which served as preliminary labels. These annotations were subsequently refined through manual enrichment by three trained annotators, following internally defined annotation guidelines covering entity relevance, semantic consistency, and contextual appropriateness. Inter-annotator agreement was measured using Cohen's κ, yielding an average score of 0.82, indicating strong consistency across annotators. In total, the final dataset comprises 110 Tamil e-books, spanning classical literature, modern fiction, essays, philosophical texts, and socio-cultural narratives. The dataset was used as a single unified corpus for all experiments, training, evaluation, and ablation studies conducted within the TamilBookAnnot framework. This dataset design ensures genre diversity, semantic richness, and sufficient linguistic variability to rigorously evaluate the proposed annotation methodology.

4 Performance Evaluation and Results

The performance of the proposed strategic Tamil e-book annotation framework, TamilBookAnnot, is validated using four primary metrics: precision, recall, accuracy, and F-measure, with False Discovery Rate (FDR) employed as an auxiliary metric. The reason for choosing these metrics in their averaged forms is that they directly quantify the relevance and correctness of the annotation results, while the FDR provides a quantitative assessment of error rates and deviations in the model. Collectively, these metrics ensure that the evaluation reflects not only the strength of correct predictions but also the robustness of the framework in minimizing errors. From Table 1, it is observed that the proposed TamilBookAnnot achieves an overall precision of 96.89%, an average recall of 97.98%, an accuracy of 97.44%, and the lowest FDR value of 0.04. These results confirm that the framework consistently produces annotations of high quality and minimal error, outperforming alternative models tested under the same experimental setup. To establish a meaningful comparison, TamilBookAnnot is baselined with three distinct frameworks, namely TamilNER, ANERTamil, and VidAnnot. Since there are limited models specifically developed for e-book annotation, the baselines are drawn from related areas such as Tamil Named Entity Recognition (NER) and video annotation frameworks. While TamilNER and ANERTamil are primarily NER-based models, their recognized entities can be used as proxies for annotation tasks. VidAnnot, originally developed for video annotation and recommendation, is adapted to the e-book domain since it uses similar strategies of annotation generation and semantic reasoning. All baseline models and TamilBookAnnot are implemented in the same environment, on the exact same dataset, with appropriate APIs for translation and metadata handling, ensuring that the comparison is balanced and fair.

From Table 1, it is evident that TamilBookAnnot yields the highest values in precision, recall, accuracy, and F-measure, along with the lowest FDR, clearly surpassing the baselines. The main reason for this consistent outperformance is the multi-layered hybridization of knowledge derivation and semantic reasoning schemes in TamilBookAnnot. Unlike baseline models that depend solely on dataset-specific knowledge, TamilBookAnnot strategically integrates auxiliary knowledge from multiple perspectives. The framework first extracts terms and categories directly in Tamil and applies TF-IDF on the Tamil book dataset to derive informative terms. These Tamil terms and categories are then contextually enriched using Tamil WordNet, ensuring structural and lexical contextualization. Additionally, TamilBookAnnot leverages the Tamil LLM 7B built on Gemma 7B, which generates Tamil text at scale, thereby contributing to rich Tamil knowledge derivation. A significant novelty lies in how Tamil terms extracted via TF-IDF are converted into English, since the structural metadata of Web 3.0 is predominantly English-based. This allows the system to access large-scale metadata repositories, which are then classified using BERT to yield atomic, manageable units. The classified metadata is subsequently translated back into Tamil via Lingvanex, ensuring that the enriched knowledge remains accessible in the target language. By following this strategy, TamilBookAnnot is able to combine high-density auxiliary knowledge from English metadata with native Tamil linguistic representations, effectively balancing global knowledge repositories with local language requirements. Further, knowledge graph generation through Google's Knowledge Graph API, combined with CYC–YAGO pipelines, provides access to community-contributed and community-verified knowledge. Translating these enriched subgraphs into Tamil adds another layer of auxiliary knowledge to the framework. This pipeline makes TamilBookAnnot distinctively rich in auxiliary knowledge density, blending formalized metadata with community-verified repositories. The inclusion of SimRank and Lloyd's Index, with differential thresholds and step-deviance measures, ensures strong quantitative semantic reasoning within the pipeline. These similarity computations allow the framework to identify cohesive and contextually relevant entities across the multilingual knowledge layers. Moreover, the use of Charged System Search (CSS) as a metaheuristic optimization algorithm, with Lloyd's Index set as the criteria function, enables the computation of the optimal solution set. This optimal set is then enriched by feedback-based reinforcement through the knowledge graph, CYC, and YAGO pipelines, ensuring continuous refinement of annotations.

The TamilNER framework, although designed as a Tamil annotator, does not perform well compared to TamilBookAnnot. Its limitations stem primarily from its high domain specificity, as it was originally designed for biomedical document annotation. TamilNER relies on an SVM classifier, along with traditional features such as bigrams, trigrams, case markers, and substring clues. While substring clues and TF-IDF scores

Table 1. Comparison of performance of the proposed TamilBookAnnot with other approaches.

Model	Average Precision%	Average Recall%	Average Accuracy%	Average F-Measure%	FDR (False Discovery Rate)
TamilNER [9]	87.87	89.84	88.86	88.84	0.13
ANERTamil [10]	90.22	93.801	92.01	91.98	0.10
VidAnnot [11]	92.44	93.89	93.17	93.16	0.08
Proposed TamilBookAnnot	**96.89**	**97.98**	**97.44**	**97.43**	**0.04**

provide lightweight auxiliary signals, they are insufficient for large-scale, heterogeneous annotation tasks. TamilNER lacks semantic reasoning capabilities and does not incorporate auxiliary knowledge from external sources. Its domain-constrained auxiliary knowledge and absence of generative or deep learning components significantly weaken its performance when benchmarked against TamilBookAnnot. The ANERTamil framework, or Automatic Tamil Entity Recognition, also lags behind. Though hybridized with language-specific features tailored for Tamil NER, it is essentially a corpus-driven system that relies heavily on handcrafted linguistic markers. Auxiliary knowledge in ANERTamil is extremely lightweight, as it lacks integration with large-scale repositories such as WordNet or knowledge graphs. Moreover, there is no incorporation of LLMs, no community-driven enrichment, and no strong quantitative semantic reasoning schemes. As a result, ANERTamil fails to scale effectively in the Web 3.0 ecosystem, producing lower recall and accuracy compared to TamilBookAnnot.

The VidAnnot framework, adapted from a video annotation model, also struggles when applied to e-book annotation. While VidAnnot is designed for semantic connotation in video lectures and recommendation tasks, it relies heavily on content-based similarity and lightweight annotation schemes. Although it can generate annotations based on video content, its auxiliary knowledge encompassment is minimal, and its semantic reasoning mechanisms are relatively weak. When adapted for e-books, these shortcomings become more apparent, leading to significant underperformance compared to TamilBookAnnot. As shown in Fig. 2, the precision versus number of recommendations distribution curve further validates these outcomes. TamilBookAnnot consistently occupies the highest position in the hierarchy, while TamilNER occupies the lowest. ANERTamil follows next, and VidAnnot occupies the penultimate position.

The superior performance of TamilBookAnnot lies in its multi-perspective auxiliary knowledge generation and integration pipeline. By combining direct Tamil term extraction, TF-IDF-driven informative terms, WordNet contextualization, Tamil LLM-based text generation, BERT-based metadata classification, and translation pipelines via Lingvanex, it successfully bridges the gap between Tamil-native knowledge and English-dominated Web 3.0 metadata. The further inclusion of community-verified knowledge repositories (via Google Knowledge Graph, CYC, and YAGO) enriches its auxiliary density, making the annotations robust, semantically cohesive, and linguistically precise. The combination of SimRank, Lloyd's Index with differential thresholds, and CSS

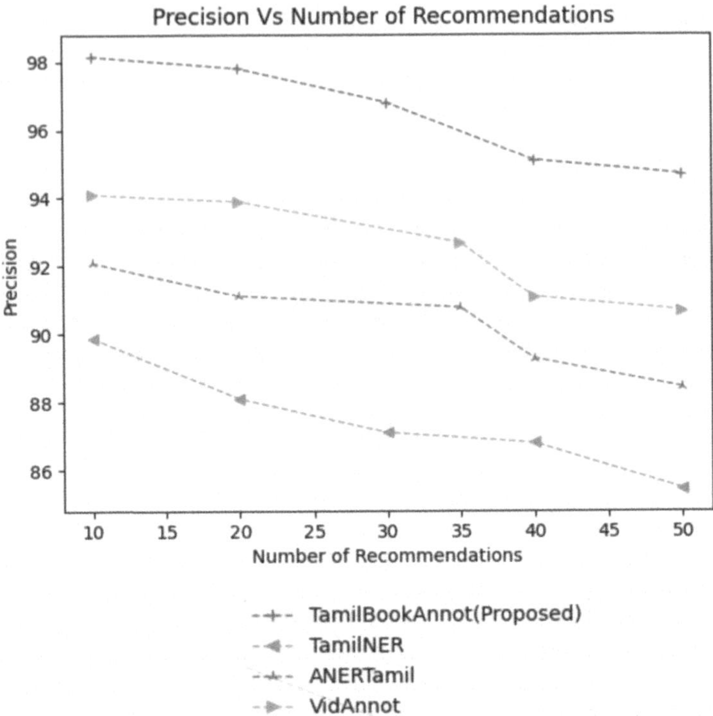

Fig. 2. Precision Percentage Vs Number of Recommendations

metaheuristic optimization ensures that the annotation process is not only accurate but also computationally optimal. By aligning auxiliary knowledge with semantic similarity measures and optimization algorithms, TamilBookAnnot achieves results as a robust e-book annotation framework.

Table 2 depicts the ablation study for the proposed TamilBookAnnot framework. When Gemma 7B is eliminated from TamilBookAnnot, it is seen that there is a 5.16% decrease in the overall accuracy, which clearly indicates that the Gemma 7B yields high-density auxiliary knowledge that is extracted from text and extended as captions for the text summaries generated by the large language model. Large language models like Gemma 7B are highly capable of yielding domain-specific, specialized entities directly relevant to the underlying Tamil literary domain. This clearly shows that, based on the dataset terms, categories, and the domain context, Gemma 7B contributes robust, highly specialized knowledge that is augmented into the model. Therefore, eliminating Gemma 7B results in a very significant 5.16% drop in overall accuracy. When Google's Knowledge Graph API is eliminated from TamilBookAnnot, there is a 3.18% decrease in the overall accuracy. Google's Knowledge Graph API houses high-density, cognitively capable knowledge graphs and subgraphs that capture human cognition from community-contributed and community-verified repositories. Since the Knowledge Graph API yields a large amount of structured entities, its removal leads to a notable loss of cognitively grounded knowledge, hence the 3.18% dip. When CYC is eliminated from the proposed

TamilBookAnnot, a 1.17% decrease in accuracy is observed. This is mainly because CYC is also a community-contributed, community-verified knowledge store repository, but the number of highly domain-specific entities coming from CYC may not be as significant for the Tamil literary domain. Therefore, the decrease remains moderate at 1.17%. Similarly, when YAGO is removed, a 1.89% decrease in accuracy is seen. As YAGO also contributes community-verified entities, although not always highly specific to the domain, its removal still affects the framework modestly. Similarly, when the BERT classifier is eliminated from the proposed TamilBookAnnot, a 4.62% decrease in overall accuracy is seen.

The primary reason is that the BERT classifier acts as a high-density metadata classifier, improving the permeability and structuring of metadata, making it atomic and capable of permeating into the proposed system as a well-organized unit. Without classification, the metadata remains haphazard, and the inclusion of well-seasoned classes is not ensured, leading to a significant accuracy decline. When metadata itself is eliminated completely, an 8.22% drop in accuracy is observed. This is because metadata directly comes from Web 3.0 structures and document contents, and removing metadata leads to an exponential loss of entities that would otherwise be integrated into the framework, drastically reducing the overall accuracy.

Table 2. Ablation study

Removed Component	Decrease in Overall Accuracy (%)
Eliminating Gemma 7B from TamilBookAnnot	5.16
Eliminating Google KG API from TamilBookAnnot	3.18
Eliminating CYC from TamilBookAnnot	1.17
Eliminating YAGO from TamilBookAnnot	1.89
Eliminating BERT Classification from TamilBookAnnot	4.62
Eliminating metadata from TamilBookAnnot	8.22
Eliminating Charged System Search from TamilBookAnnot	5.42
Eliminating SimRank from TamilBookAnnot	2.02
Eliminating Lloyds Index from TamilBookAnnot	2.88

Subsequently, when CSS (Charged System Search) is eliminated, an overall 5.42% decrease in accuracy is seen, which clearly indicates that a concept-inspired meta-heuristic optimization is required to transform the initial solution into a highly optimal solution through the criteria function. Without CSS, the optimization step weakens, leading to reduced performance. When SimRank is eliminated, a 2.02% decrease is seen because quantitative reasoning certainly plays a role in identifying meaningful semantic relationships. Similarly, with Lloyd's Index, a 2.88% decrease is seen, again showing that quantitative semantic reasoning is essential. Without the presence of SimRank and Lloyd's Index, there is a clear dip in overall accuracy. Although the TamilBookAnnot framework employs a broad components: TF-IDF, WordNet, Gemma 7B, Google's Knowledge Graph API, CYC, YAGO, SimRank, Lloyd's Index, and CSS, the pipeline is a inference-oriented, making it computationally feasible despite its structural depth. Only the Random Forest classifier undergoes training, and it remains a lightweight ML model with minimal computational footprint. All other modules contribute through incremental semantic filtering, reasoning, and knowledge aggregation, rather than repeated end-to-end learning cycles. These observations confirm that while each module meaningfully enriches semantic robustness, the overall framework remains computationally practical, deployable, and not resource-intensive.

5 Conclusion

This paper presents TamilBookAnnot, a strategic e-book annotation framework for Tamil that integrates semantic artificial intelligence with large language models, generative enrichment, quantitative semantic reasoning, and metaheuristic optimization. The framework addresses the cognitive gap between structured knowledge representations prevalent in Web 3.0 and the semantic processing requirements of regional language content. Incremental auxiliary knowledge aggregation through Tamil WordNet and generative modeling using Gemma 7B enables linguistic coherence and contextual depth, while metadata generation and classification facilitate structured semantic representations. Quantitative semantic reasoning through similarity measures and optimization via Charged System Search enhance convergence toward highly relevant annotation sets by transforming feasible solutions into optimally refined outputs. The integration of semantic intelligence, generative artificial intelligence, and concept-driven optimization establishes a scalable and extensible foundation for Tamil digital knowledge organization. Experimental evaluation demonstrates consistent performance gains across multiple metrics, indicating the framework's suitability for large-scale Tamil e-book annotation. Future extensions may explore multilingual adaptation, cross-domain semantic interoperability, and low-resource language integration within emerging Web 3.0 knowledge ecosystems.

References

1. Lacic, E., et al.: Evaluating tag recommendations for e-book annotation using a semantic similarity metric. arXiv preprint arXiv:1908.04042 (2019)
2. Hwang, H.K., Ronchetti, M.: Q-Book: multimedia annotation tool for e-books. In: EdMedia+ Innovate Learning, pp. 2543–2551. Association for the Advancement of Computing in Education (AACE) (2014)
3. Ma, B., Chen, L.: A Framework for Constructing Concept Maps from E-Books Using Large Language Models: Challenges and Future Directions
4. Chang, T.P., Chen, H.M., Chen, J.Q.: A deep learning co-training framework for e-book classification. In: 2020 International Symposium on Computer, Consumer and Control (IS3C), pp. 376–379. IEEE (2020)

5. Indranandita, A., Susanto, B., Rahmat, A.: Journal classification and search system using the naive bayes method and vector space model. Inform. J. **4**(2) (2011)
6. Brouard, C.: Document classification by computing an echo in a very simple neural network. In: 2012 IEEE 24th International Conference on Tools with Artificial Intelligence, vol. 1, pp. 735–741. IEEE (2012)
7. Zhang, W., Yoshida, T., Tang, X.: A comparative study of TF* IDF, LSI and multi-words for text classification. Expert Syst. Appl. **38**(3), 2758–2765 (2011)
8. Devika, P., Milton, A.: Book recommendation system: reviewing different techniques and approaches. Int. J. Digit. Libr. **25**(4), 803–824 (2024)
9. Antony, J.B., Mahalakshmi, G.S.: Named entity recognition for Tamil biomedical documents. In: 2014 International Conference on Circuits, Power and Computing Technologies [ICCPCT-2014], pp. 1571–1577. IEEE (2014)
10. Srinivasan, R., Subalalitha, C.N.: Automated named entity recognition from Tamil documents. In: 2019 IEEE 1st International Conference on Energy, Systems and Information Processing (ICESIP), pp. 1–5. IEEE (2019)
11. Dias, L.L., Barrére, E., de Souza, J.F.: The impact of semantic annotation techniques on content-based video lecture recommendation. J. Inf. Sci. **47**(6), 740–752 (2021)

AssameseDocRec: A Strategic Framework for Assamese Document Recommendation Integrating Generative Semantic Intelligence and Quantitative Semantic Reasoning

Anubrat Bora[1], Gerard Deepak[2(✉)], and Samiksha Shukla[2]

[1] School of Computer Science and Engineering, Manipal Institute of Technology Bengaluru, Bengaluru, India

[2] Department of Computer Science and Engineering (Cyber Security), Dayananda Sagar Academy of Technology and Management, Bangalore, India

gerard.deepak.christuni@gmail.com

Abstract. This study introduces a strategic framework for Assamese document recommendation (AssameseDocRec), with emphasis on the language's heritage and literature. The framework mainly integrates generative and semantic intelligence to address the limitations of conventional keyword-based or ranking approaches. IndicBERT is used for the document classification task, while the TF-IDF and AxomiyaBERTa support metadata generation and auxiliary knowledge enrichment. Semantic reasoning is achieved through CoSimRank and the Petraitis index, with Invasive Weed Optimization applied to refine the recommendation set. It outperforms three baseline models by achieving 96.89% precision, 97.87% recall, 97.38% accuracy, and a 97.38% F-measure, while also maintaining a false discovery rate of only 0.04. These results empirically demonstrate the robustness and effectiveness of the framework. Thus, the system enhances the visibility of Assamese heritage and literature within emerging Web 3.0 ecosystems by unifying various generative AI, semantic reasoning, and optimization techniques.

Keywords: Document Recommendation · Semantic Intelligence · Generative AI · Quantitative Reasoning

1 Introduction

The growth of digital repositories has made document recommendation a central task for users seeking relevant knowledge in large corpora. Traditional approaches based on keyword matching or collaborative filtering are often inadequate when faced with heterogeneous datasets, multilingual corpora, and code-mixed inputs. The demand today is for systems that can integrate semantics, context, and generative intelligence to guide users toward meaningful information.

In the era of Generative AI, recommendation extends beyond retrieval. Generative models can summarize, rephrase, and adapt documents to user intent, thereby improving

© The Author(s), under exclusive license to Springer Nature Switzerland AG 2026
A. Kannan et al. (Eds.): ADCOM 2025, CCIS 2947, pp. 160–172, 2026.
https://doi.org/10.1007/978-3-032-26269-1_11

both accessibility and comprehension. This shift has redefined recommendation as a generative-semantic task rather than a simple ranking problem. At the same time, Web 3.0 environments emphasize decentralized metadata, community-driven knowledge graphs, and interoperability across platforms. In such a setting, recommendation frameworks must integrate semantic intelligence with generative capabilities while remaining flexible enough to interact with large-scale metadata systems.

For Assamese, the need is especially pressing. Assamese is an original and historically rich language of Northeastern India, spoken by millions, yet its presence on the digital web is limited. The lack of Assamese metadata, coupled with minimal indexing support, reduces its visibility in global knowledge networks. Web 3.0 offers an opportunity to reverse this scarcity by embedding Assamese documents into distributed metadata systems, but only if robust frameworks exist to process, translate, and recommend them effectively.

Thus, there is a requirement for an Assamese document recommendation framework that not only addresses low-resource challenges but also integrates Generative AI, Semantic AI, and quantitative semantic reasoning. Such a framework can bridge cultural knowledge gaps, preserve linguistic heritage, and ensure that Assamese literature and heritage documents are discoverable in modern digital ecosystems.

Motivation: Domains such as heritage and literature have long been neglected in the design of mainstream recommendation systems. Assamese, however, is a language rich in culture, heritage, and literary tradition, with a vast number of documents rooted in its history and identity. Despite this richness, there is no high-performing recommendation or indexing framework available for Assamese, which has limited the accessibility of its knowledge base. With Web 3.0 coming up, Assamese documents are starting to show up in community archives and decentralized platforms. But their presence is scattered, not well indexed, and mostly hidden in larger systems. To make sure this knowledge is preserved and easy to reach, the role of Assamese in Web 3.0 has to be made stronger. As Web 3.0 grows out of Web 2.0, in line with Sir Tim Berners-Lee's vision of a semantic and connected web, the need for Assamese to have a place in this space is clear. What is required is a mix of Generative AI and Semantic Intelligence, using semantic reasoning to support a recommendation framework that works for Assamese. The work in this paper is aimed at that need, focusing on Assamese heritage and literature as the chosen domain.

Contributions: This paper proposes a strategic framework for Assamese document recommendation, with a special focus on heritage and literature as the target domain. The framework uses IndicBERT to classify the dataset in Assamese, ensuring that documents are handled in their native form rather than through only translation. Metadata generation is carried out from entities extracted using TF-IDF, which are then passed to AxomiyaBERTa, an Assamese language model, to produce dense auxiliary knowledge. This makes both the dataset and the user query more informative at different stages of processing. For semantic reasoning and learning, the framework applies the Petraitis Index and CoSimRank with empirically chosen thresholds and study-level measures to achieve quantitative semantic reasoning. In parallel, LDA is used for topic modeling, and the query is enriched through Google's Knowledge Graph API. The enriched entities, along with the classified dataset instances, are aggregated. The initial solution set produced by these steps is then refined using Invasive Weed Optimization (IWO), with the Petraitis Index guiding the selection of optimal solutions. This pruning ensures that the recommendation set is both comprehensive and precise. The novelty of this framework lies in its integration of generative and semantic intelligence. AxomiyaBERTa generates summaries and metadata, IndicBERT classifies Assamese texts, LDA provides topic structure, while Petraitis Index and CoSimRank offer quantitative reasoning. Together, these elements hybridize generative AI with semantic intelligence to build a robust recommendation pipeline for Assamese heritage and literature. Each component of the framework fulfills a specific functional role. TF-IDF provides lexical grounding and candidate term extraction, while AxomiyaBERTa enables semantic abstraction through contextual summarization. IndicBERT strengthens multilingual classification, and Latent Dirichlet Allocation (LDA) captures interpretable topic structures. CoSimRank and the Petraitis Index jointly provide quantitative semantic similarity, while Invasive Weed Optimization iteratively refines candidate recommendations. Together, these modules integrate heterogeneous semantic signals within a unified recommendation pipeline.

2 Related Work

Work in Assamese NLP has moved in small but important steps, often beginning with foundational tasks such as part-of-speech tagging. Barman et al. [1] made one of the first systematic attempts, experimenting with CRF++ and fnTBL. The accuracy they achieved was limited, but the study showed that statistical sequence models could be adapted to Assamese despite its irregular grammar. Soon after, Daimary et al. [2] applied hidden Markov models and obtained more stable results, though they too pointed out that the absence of large, annotated corpora remained a serious bottleneck. The field saw a stronger breakthrough with Pathak et al. [3], who introduced AsPOS, a BiLSTM-CRF system that incorporated multilingual embeddings from MuRIL and Flair. By borrowing strength from richer languages, their approach pushed performance to a new level and made clear the value of transfer learning for Assamese.

Morphological analysis has also received attention. Rahman and Sarma [4] worked with the Apertium platform to build one of the earliest Assamese morphological analyzers. Their rule-based design tackled inflectional patterns that had long been seen as too complex for automation, and in doing so they created a tool still relevant for translation and other text-processing tasks. In a similar spirit of building resources, Pathak et al. [5] compiled AsNER, a dataset for named entity recognition, and tested baseline models. This was an important step because it filled a gap in resources that had kept Assamese NER work lagging behind other Indian languages.

Sentiment analysis has become another focus area. Das and Singh [6] explored classification of Assamese news and social media content. Their study made clear how noisy, user-generated data can complicate NLP tasks, especially in a language with few existing tools. The same authors [7] later experimented with multimodal designs, combining text, captions, and images through late fusion. This move beyond text-only processing opened a new line of inquiry, where sentiment prediction draws on visual as well as linguistic cues. Expanding on this direction, Das et al. [8] proposed a multi-stage multimodal framework that further refined how these signals could be combined, making a case for richer feature integration even in low-resource settings.

The development of parallel resources has been just as critical. Baruah et al. [9] built an Assamese English bilingual corpus that made statistical machine translation experiments possible and provided data for training other bilingual models. More recently, Tamang and Bora [10] created a centralized repository that brings together multiple Assamese datasets across sentiment analysis, classification, and translation. By consolidating scattered efforts, they offered researchers a practical foundation for building and testing new systems.

These studies show serious growth. Early rule-based and statistical models established baselines which are also enriched with multilingual embeddings have truly brought measurable gains; and resource development has ensured that the language is not left entirely behind in the wave of modern NLP.

3 Methodology

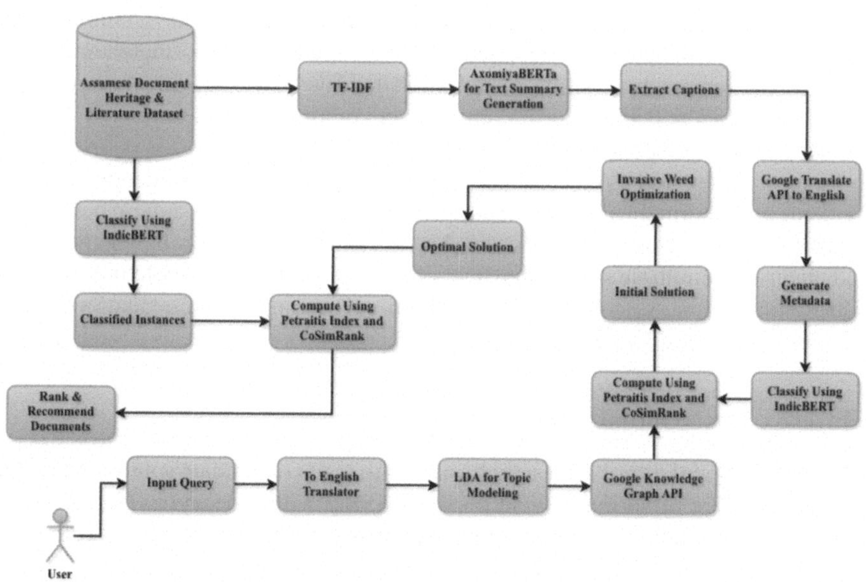

Fig. 1. Architecture of the Proposed AssameseDocRec Framework

Figure 1 illustrates the proposed system architecture for the strategic framework of Assamese document recommendation powered by generative semantic artificial intelligence. The Assamese Document Heritage and Literature Dataset serve as the primary knowledge source, along with the user's query, which may be in Assamese, English, or a mixture of both.

The first step involves extracting informative terms from the dataset using term frequency–inverse document frequency (TF-IDF). Both frequent and rare terms in the corpus are considered, and the resulting terms are processed using AxomiyaBERTa, a BERT variant specifically built for Assamese, to generate text summaries. From these summaries, captions are extracted and translated from Assamese to English using Google's Translation API. Translation is necessary because most metadata resources (Web 3.0 subscription metadata, totaling over 3.02 million records) are predominantly in English, whereas Assamese metadata remains scarce and poorly supported by standard tools. The translated summaries are then passed to DSpace, the chosen tool for generating structured metadata.

Parallel to this, the dataset undergoes classification using IndicBERT, which is capable of handling Assamese classification tasks without handcrafted feature engineering. User queries are preprocessed using NLTK and also translated into English, after which Latent Dirichlet Allocation (LDA) is applied for topic modeling. LDA produces highly relevant but limited topic entities, which are further enriched through Google's Knowledge Graph API to incorporate community-verified knowledge. The resulting entities are then translated back into Assamese and classified using IndicBERT.

The entities derived from LDA, IndicBERT, and metadata generation are aggregated and filtered using a two-step semantic thresholding process: first at 0.75 similarity, and later through invasive weed optimization (IWO), a nature-inspired metaheuristic that refines the solution set by visiting each entity iteratively. To balance stringency, the Petraitis Index is set at 0.12, ensuring optimal yet comprehensive solutions.

Finally, the optimized solution set is compared with the classified dataset entities using Petraitis Index and CoSimRank (threshold 0.8) to measure semantic closeness. The ranked results form the recommendation output, focusing on Assamese heritage and literature documents. This framework goes beyond simple translation-based approaches, instead integrating multiple knowledge repositories, semantic topic modeling, and metaheuristic optimization. As a result, it delivers high-performing recommendations tailored for Assamese literature and heritage resources.

3.1 Invasive Weed Optimization

IWO is a simple population-based search: start with seeds (candidate solutions), let fitter seeds produce more offspring, spread those offspring with a decaying Gaussian, then apply competitive exclusion to keep only the strong. Over iterations, the colony fills good regions but still explores because the spread starts wide and narrows over time. That fits our problem: lots of candidate entities after enrichment, but we want a compact, diverse set that still scores well.

We initialize with all entities that pass an initial semantic threshold. Reproduction is proportional to a fitness signal. Spatial dispersion uses a standard deviation that shrinks each iteration, so early steps explore, and later steps refine. Competitive exclusion caps population size, dropping near-duplicates. This keeps runtime reasonable without overfitting the first good cluster.

Compared with full evolutionary toolkits, IWO is light to tune and robust when the objective is bumpy (our case: multiple noisy signals). Recent work still extends it (event-triggered seeding, chaotic dispersion), but the core is stable and easy to explain in a methods section.

3.2 IndicBERT

IndicBERT is a multilingual ALBERT trained on ~9B Indic tokens across 12 languages, including Assamese. Because it's lightweight and covers Assamese directly, it's a natural classifier/encoder in our stack (no handcrafted features). In practice, this helps on mixed-script/mixed-language inputs and keeps inference costs low.

3.3 AxomiyaBERTa

AxomiyaBERTa is a monolingual Assamese Transformer trained with masked language modeling, designed specifically for low-resource settings. By focusing on Assamese morphology and syntax, it achieves strong results on token-level tasks such as NER, making it a reliable choice for Assamese-centric representation learning in summarization and generation.

AxomiyaBERTa is used in a pretrained configuration to perform abstractive summarization. Documents are encoded using contextual embeddings, and salient sentences are selected based on semantic similarity to the global document representation. Task-specific fine-tuning is not performed due to the limited availability of annotated Assamese summarization datasets. The generated summaries are subsequently used for metadata enrichment and entity extraction.

3.4 Latent Dirichlet Allocation

LDA (Latent Dirichlet Allocation) is a generative probabilistic model that represents each document as a mixture of latent topics, where each topic is a distribution over words. It provides interpretable, high-precision topic/entity representations that support document ranking and summarization.

3.5 Term Frequency Inverse Document Frequency

TF-IDF (Term Frequency Inverse Document Frequency) is a statistical measure that evaluates the importance of a word in a document relative to a corpus. It increases with term frequency within a document but is offset by how commonly the term appears across the collection, thus balancing relevance and uniqueness.

3.6 CoSimRank

CoSimRank is a graph-based measure of similarity that determines the equivalence of two entities through measuring the similarity of their respective neighborhoods. The algorithm represents a combination of SimRank and the cosine similarity, which makes it more stable and efficient in terms of calculations especially when dealing with large graph structures. As a result, CoSimRank has been extensively used in the field of link analysis, recommendation systems, as well as assessing semantic similarity.

3.7 Petraitis Index

The Petraitis Index is a score that is used to measure the difference impact of an intervention in comparison to a control. It expresses the size of the effect and introduces a factor of inherent variability of the dataset. A high index is an indication of a gain in the response over the control and a lower index is a gain in the response. This index is commonly used in ecological as well as biological studies in order to allow fair comparison of various treatment modalities.

Even though initially suggested as part of the ecological statistics, the Petraitis Index is used to measure divergence between distributions. In the given framework, document entities and query entities are defined as the semantic distributions. The index, therefore, measures difference between semantic populations thus allowing similarity estimation on the distribution level as opposed to using pairwise cosine similarity. This adaption allows strong comparison of semantic network in low-resource multilingual settings and supplements graph similarity measurements conducted through CoSimRank.

4 Results and Performance Evaluation

Hybridization of four different datasets was carried out. The FutureBeeAI English–Assamese parallel corpus for the cultural domain dataset [11], the Assamese Corpus 2022 dataset (B2FIND) [12], the Work With Data 2025 dataset of books about Assam (social life and customs, fiction) [13], and the Work With Data 2025 dataset of books about Assam (description and travel) [14] were included. This was strategically designed in such a way that the dataset, in its entirety, was subjected to extraction of terms and captions, which were then converted into Assamese through translation APIs. Using these transliterated entities, customized crawlers queried Assamese language literature books, retrieving all relevant keywords and key terms in Assamese. From this, a single large corpus in Assamese was formed, with an English background and common captions in both Assamese and English. A consolidated Assamese corpus on travel, culture, heritage, social life, and related domains was formalized, focusing specifically on Assamese. If the dataset was bilingual or not originally in Assamese, only the terms were used for crawling, and then a single large dataset was formalized on which the expansions were carried out.

Ground-truth relevance labels were also acquired through manual annotation. In accordance with the semantic correspondence to the Assamese heritage and literary topics, each query was assigned the relevant or non-relevant category of documents. Three independent evaluators with academic backgrounds in computer science and linguistics were used in the annotation process and were fluent in Assamese. Disagreements were resolved through majority voting. A sample consisting of 120 user queries, both Assamese and English as well as code-mixed queries, was tested. The combined corpus was divided into 70% training set, 15% validation set and 15%test set. The parts of the classification used in the IndicBERT were trained using the training part, and the testing of the performance of the recommendations was restricted to the held-out testing set. The metrics used were precision, recall, accuracy, F - score and false discovery rate (FDR), as based on comparison between the retrieved recommendations and the annotated ground -truth relevance marks.

The resulting consolidated corpus with 18,742 documents is based on four data collections on Assamese related to culture, heritage, literature, travel and social life. After deduplication and preprocessing, the corpus contained 14,306 Assamese-majority documents and 4,436 bilingual Assamese-English documents. Preprocessing consisted of Unicode normalization, elimination of extraneous symbols, removal of stop-words, segmentation of sentences and translation alignment verification. Documents that included less than 50 tokens were not used to ensure semantic completeness.

The performance of the proposed AssameseDocRec framework, which integrates generative AI with semantic intelligence for Assamese heritage and literature documents, has been evaluated using Precision, Recall, Accuracy, F-measure, and False Discovery Rate (FDR). Precision, Recall, Accuracy, and F-measure quantify the relevance and reliability of the recommendations, while FDR captures the degree of deviation or error. Together, these measures provide a balanced view of both positive performance and negative distribution analysis. The proposed framework achieves 96.89% Precision, 97.87% Recall, 97.38% Accuracy, and 97.38% F-measure, with an FDR of 0.04. These results demonstrate its ability to maintain high relevance while minimizing erroneous recommendations (Table 1).

Table 1. Comparison of Performance of the proposed AssameseDocRec with other approaches

Model	Average Precision %	Average Recall %	Average Accuracy %	Average F-Measure %	FDR
MFDCR [15]	88.57	90.94	90.76	89.74	0.12
DRDO [16]	90.19	92.08	91.14	91.13	0.10
LDRS [17]	93.14	94.89	94.02	94.01	0.07
Proposed AssameseDocRec	96.89	97.87	97.38	97.38	0.04

To benchmark performance, AssameseDocRec was compared against three baseline systems: MFDCR [15], DRDO [16], and LDRS [17]. MFDCR [15] attained 88.57% Precision, 90.94% Recall, and the lowest Accuracy and F-measure, with the highest FDR of 0.12. Its reliance on probabilistic matrix factorization without auxiliary knowledge integration led to underfitting and poor semantic reasoning capability. DRDO [16] yielded 90.19% Precision, 92.08% Recall, and a moderate FDR of 0.10. Although it uses ontology-based similarity and TF-IDF for term weighting, it suffers from low-density auxiliary knowledge augmentation and lacks metaheuristic optimization, leading to weak semantic reasoning. LDRS [17] performed better than the other two baselines, with 93.14% Precision, 94.89% Recall, and an FDR of 0.07. However, being highly domain-specific (originally legal texts translated into Assamese) and limited to clustering with pairwise similarity, it fails to incorporate incremental auxiliary knowledge or advanced semantic reasoning.

Overall, AssameseDocRec surpasses all baselines in every metric. Its superior performance stems from auxiliary knowledge generation and augmentation through TF-IDF,

AxomiyaBERTa, and Knowledge Graph APIs, along with metadata enrichment using translation pipelines and IndicBERT classification. Quantitative semantic reasoning is achieved using CoSimRank with thresholds and step deviance measures, while meta-heuristic optimization is performed through invasive weed optimization with Petraitis Index. The framework also supports multilingual queries, reducing the cognitive gap between English, Assamese, and mixed-language inputs (Table 2).

Table 2. Precision Percentage vs. No. of recommendations

No. of Recommendations	MFDCR [15]	DRDO [16]	LDRS [17]	Proposed AssameseDocRec
10	90.09	92.09	95.19	98.49
20	89.19	91.10	94.98	97.87
30	88.89	90.76	93.67	96.88
40	87.78	89.87	92.08	95.79
50	86.47	88.42	91.87	94.39

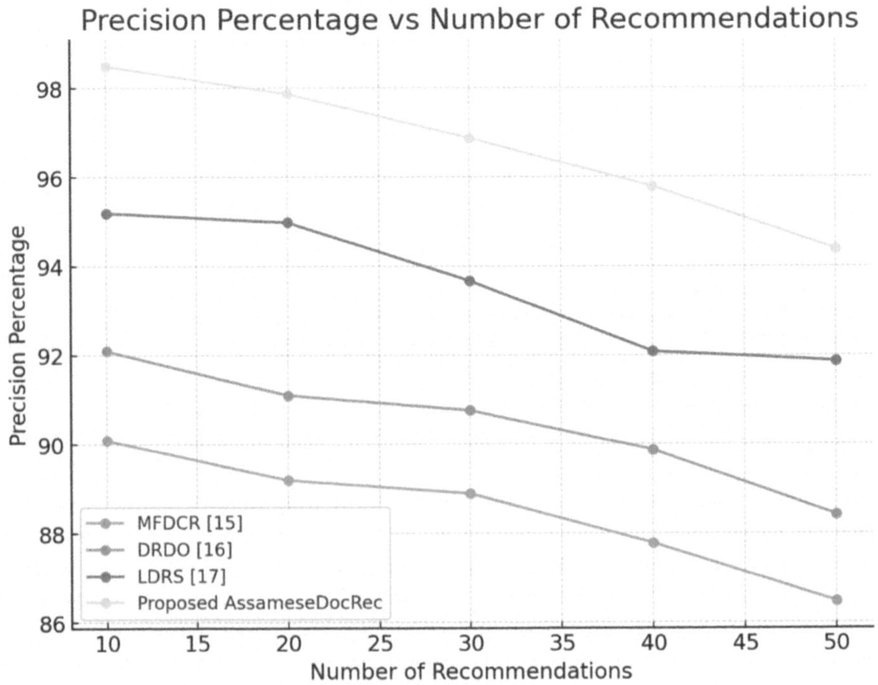

Fig. 2. Precision Percentage vs. No. of Recommendations

Figure 2 depicts the distribution of recommendations across models, where AssameseDocRec consistently occupies the top hierarchy with the widest spread of accurate recommendations. MFDCR [15] is positioned at the bottom due to its poor semantic capacity, while DRDO [16] and LDRS [17] occupy intermediate levels. LDRS, despite being stronger than the other baselines, remains below AssameseDocRec because of its lack of incremental knowledge aggregation.

In summary, the proposed framework emerges as a high-performing model for Assamese document recommendation, outperforming established baselines by integrating generative AI, semantic intelligence, and optimization strategies in a unified pipeline.

Table 3. Ablation study showing the decrease in F-measure when each component is removed from the proposed framework.

Technique Variation	Overall % Decrease in F-Measure	Resultant F-Measure
Proposed Framework minus TF-IDF	3.12%	94.26
Proposed Framework minus AxomiyaBERTa	5.63%	91.75
Proposed Framework minus Metadata	9.21%	88.17
Proposed Framework minus IndicBERT	6.12%	91.26
Proposed Framework minus Petraitis Index	3.12%	94.26
Proposed Framework minus CoSimRank	2.82%	94.56
Proposed Framework minus LDA	1.18%	96.20
Proposed Framework minus IWO	8.12%	89.26

Table 3 depicts the ablation study, which assesses the value provided by each component in the proposed framework by removing each module in a systematic manner and monitored the resulting decrease in F -measure. The exclusion of the TF-IDF module led to 3.12% decrease in F-measure. The words that are found in the corpus frequently or occur infrequently but are contextually important are identified in the TF-IDF and its removal limits the provision of significant lexical units, which reduces performance. The large-scale textual model used to produce high-quality summaries and extract entities of the caption level, which does not include AxomiyaBERTa, resulted in a decrease of 5.63%. Being a generative transformer-based model, AxomiyaBERTa offers a strong contextual semantics, in the absence of which the quality of the summary improves as well as the semantic richness of the extracted entities. Metadata integration, which utilizes Web 3.0 structural content, was the largest performance faller at 9.21%. Metadata provisions of knowledge and contextual base that cannot be overlooked in determining entity relevance; otherwise, major semantic cues are lost. The removal of IndicBERT, which gives the contextual embeddings in the Indian language, led to a 6.12% reduction. IndicBERT maintains linguistic compliance and contextualisation of Assamese and heritage datasets; the lack of it undermines language adaptation. Deleting the Petraitis Index brought out a 3.12% decrease whilst deleting CoSimRank brought out a 2.82% decrease. The two are measures of semantic similarity that facilitate quantitative semantic reasoning and ranking of relevance of entities. The omission of Latent Dirichlet Allocation (LDA) which amplifies queries with background knowledge decreased the F-measure by 1.18%, suggesting that it adds context specificity. Lastly, when the Invasive Weed Optimization (IWO) module was removed this resulted in a decrease of 8.12%. Being a meta-heuristic optimiser, IWO finds the best solutions in terms of semantics, without which the ability of the framework to refine and optimise entity selection is curtailed. The reactiveness of the suggested framework is not just in the use of established modules but, also, in how the components are altered, integrated and restructured to fit Assamese heritage and literary information. Even though direct models that include IndicBERT are followed, the compatibility with the domain is strictly considered with the linguistic characteristics of Assamese. Similarly the modules like TF-IDF and IWO serve a background purpose; the major innovation is the addition of the Petraitis Index as an optimisation measure in IWO. The framework also includes hybrid semantic reasoning mechanism which combines the Petraitis Index and CoSimRank with empirically determined thresholds and deviation measures. In this case, the Petraitis Index is applied both as a measure of semantic similarity and as an optimisation factor, and CoSimRank gives entities contextual correlation with one another. The other significant innovation is the introduction of the metadata used to overcome the cognitive disjunction that exists between Web 3.0 structural content and the suggested entity-extraction pipeline. Extracted entities are embedded into a structural semantic environment through metadata. Multilingual contextualisation with the help of the AxomiyaBERTa and a translation pipeline in support are used to strengthen linguistic adaptation. An API that reduces semantic deviations in the process of translation is Google KnowledgeGraph API that provides community-verified, structurally consistent entities. Therefore, the proposed structure has afforded the novelty in its multilingual contextualisation synthesis, metadata-based structural grounding, hybrid semantic similarity measures, and optimisation-centric extraction workflow.

5 Conclusion

This paper has proposed a strategic framework for Assamese document recommendation, with heritage and literature as the primary domain of focus. The framework employs IndicBERT to classify Assamese documents, while TF-IDF and AxomiyaBERTa are used for metadata generation and dense knowledge augmentation, making both the dataset and user queries more informative at multiple stages.

Semantic reasoning is carried out using the Petraitis Index and CoSimRank, with empirically chosen thresholds to maintain quantitative rigor. The initial solution set is refined through Invasive Weed Optimization (IWO), which yields near-optimal solutions by systematically pruning noisy entities. Metadata classification with IndicBERT and enrichment through Google's Knowledge Graph API add further depth to the recommendations. The performance evaluation shows that the proposed framework outperforms existing models such as MFDCR [15], DRDO [16], and LDRS [17]. It achieves an average precision of 96.89%, recall of 97.87%, accuracy of 97.38%, and an F-measure of 97.38%, while maintaining a very low false discovery rate (FDR) of 0.04. In comparison, the best baseline (LDRS) reached only 93.14% precision, 94.89% recall, 94.02% accuracy, 94.01% F-measure, and 0.07 FDR. Even as the number of recommendations increases, the proposed system sustains high precision, staying above 94% at 50 recommendations, whereas baselines drop more steeply.

These results confirm that the framework is not only effective but also robust across varying recommendation sizes. By combining Generative AI, Semantic AI, and quantitative reasoning, the system establishes itself as one of the first high-performing frameworks for Assamese document recommendation, directly addressing the needs of Web 3.0 knowledge ecosystems. It ensures that Assamese heritage and literature are more visible, better preserved, and easily accessible in the digital era.

References

1. Barman, U., Sarmah, D., Sarma, K.K.: POS tagging of assamese language and performance analysis of CRF++ and fnTBL approaches. In: Proceedings of ICACCI (2013)
2. Daimary, D., Sharma, U., Sarma, K.K.: Development of part of speech tagger for assamese using hidden Markov model. International Journal of Computer Applications (2018)
3. Pathak, A., Nandi, P., Sarmah, D.: AsPOS: Assamese Part of Speech Tagger using Deep Learning Approach. arXiv preprint arXiv:2212.07043 (2022)
4. Rahman, M., Sarma, K.K.: An Implementation of Apertium Based Assamese Morphological Analyzer. arXiv preprint arXiv:1503.03989 (2015)
5. Pathak, A., Nandi, P., Sarmah, D.: AsNER: Annotated Dataset and Baseline for Assamese Named Entity Recognition. arXiv preprint arXiv:2207.03422 (2022)
6. Das, R., Singh, T.D.: Sentiment analysis in Assamese: a resource-scarce language. In: Proceedings of ICACCI (2022)
7. Das, R., Singh, T.D.: Image-text multimodal sentiment analysis framework of assamese news articles using late fusion. In: Proceedings of IHCI (2023)
8. Das, R., Singh, T.D.: A Multi-Stage Multimodal Framework for Sentiment Analysis of Assamese in a Low-Resource Setting. OpenReview preprint (2022)
9. Baruah, B., Barman, U., Sarma, K.K.: Development of English-assamese bilingual corpus for statistical machine translation. In: Proceedings of IJCNLP Workshop on Asian Language Resources (2016)
10. Tamang, S., Bora, A.: Enhancing Assamese NLP Capabilities: Introducing a Centralized Dataset Repository. arXiv preprint arXiv:2410.11291 (2024)
11. FutureBee AI. English-Assamese Parallel Corpus for the Culture Domain (2022). https://www.futurebeeai.com/dataset/parallel-corpora/assamese-english-translated-parallel-corpus-for-culture-domain
12. Assamese Corpus - Dataset - B2FIND (2022). https://b2find.eudat.eu/dataset/286fff71-a030-5743-93b1-40d3bdf1a455

13. Work With Data. Dataset of books about Assam (India)-Social life and customs-Fiction (2025). https://www.workwithdata.com/datasets/books?f=1&fcol0=j0-book_subject&fop0=%3D&fval0=Assam+%28India%29-Social+life+and+customs-Fiction&j=1&j0=book_subjects
14. Work With Data. Dataset of books about Assam (India)-Description and travel (2025). https://www.workwithdata.com/datasets/books?f=1&fcol0=j0-book_subject&fop0=%3D&fval0=Assam+%28India%29-Description+and+travel&j=1&j0=book_subjects
15. Kim, D., Park, C., Oh, J., Lee, S., Yu, H.: Convolutional matrix factorization for document context-aware recommendation. In: Proceedings of the 10th ACM Conference on Recommender Systems (RecSys'16). Association for Computing Machinery, New York, NY, USA, pp. 233–240 (2016). https://doi.org/10.1145/2959100.2959165
16. Vences Nava, R., Menendez Dominguez, V.H., Gomez Montalvo, J.: A document recommendation system using a document-similarity ontology. In: IEEE Latin America Transactions, vol. 14, no. 7, pp. 3329–3334 (2016). https://doi.org/10.1109/TLA.2016.7587638
17. Dhanani, J., Mehta, R., Rana, D.: Legal document recommendation system: a cluster based pairwise similarity computation. J. Intell. Fuzzy Syst. **41**(5), 5497–5509 (2021). https://doi.org/10.3233/JIFS-189871

AssameseVidRec: A Strategic Framework for Assamese Video Recommendation Integrating Generative Semantic Intelligence

Anubrat Bora[1] and Gerard Deepak[2](✉)

[1] School of Computer Engineering, Department of Computer Science and Engineering (Cyber Security), Department of Data Science and Engineering, Manipal Institute of Technology Bengaluru, Manchenahalli, India

[2] Dayananda Sagar Academy of Technology and Management, Bangalore, India
gerard.deepak.christuni@gmail.com

Abstract. This paper proposes a strategic video recommendation framework focusing on the hybridization of generative AI and intelligent upstream knowledge aggregation. Through this approach, a Web 3.0 compliant Assamese video recommendation system is developed that encompasses bilingual query processing in both Assamese and English. The framework ensures query word enrichment by drawing from multilingual data sources and handles translation tasks effectively. Entities are extracted from the dataset in Assamese and transformed into English, so that metadata can be generated under Web 3.0 standards using ontology-based classification and autoencoders. The framework further introduces an oriented computation paradigm where the model applies SimRank with black label divisions and differential threshold measures for semantic reasoning. Quantitative filtering is performed to refine entities, with collocation using SimRank as an objective function. This is optimized through a nature-inspired optimization technique, thus enhancing performance. Experimental evaluation shows an overall high recall and accuracy, making the framework a strong model for video recommendation, with Assamese as the chosen language of focus.

Keywords: Video Recommendation · Semantics · Web 3.0

1 Introduction

The idea of Web 3.0 has brought forward the requirement for systems that are not only interactive but also capable of understanding meaning and context. A central part of this vision is video recommendation. In a multilingual country like India, it is important that such systems do not remain limited to global languages alone. Assamese, which is one of the major regional languages of the Northeast and also recognized officially in India, is a good example of why this concern matters. At present, the use of Assamese in digital and semantic web research is limited. If the next stage of the web is to be inclusive, then languages like Assamese must be placed at the core of development rather than at the margins.

© The Author(s), under exclusive license to Springer Nature Switzerland AG 2026
A. Kannan et al. (Eds.): ADCOM 2025, CCIS 2947, pp. 173–182, 2026.
https://doi.org/10.1007/978-3-032-26269-1_12

Video recommendation frameworks provide an important entry point to this discussion. Most existing systems depend on metadata or simple user histories, which do not fully capture the meaning or cultural background of the content. With the shift towards Web 3.0, recommendation methods are expected to use semantics, annotations, and contextual features so that they can provide more accurate and personalized suggestions. For a language such as Assamese, this is not only a technical matter but also one that helps in the promotion of its cultural and digital presence.

Recent work in the field shows that machine learning, deep learning, and even generative AI methods are now being applied to recommendation problems. Research on annotation-driven models and semantic frameworks indicates that combining these approaches can create systems better suited for multilingual use. Building on these directions, a framework that highlights Assamese content within a broader multilingual setting can contribute both to academic research and to the social goal of digital inclusivity.

Motivation: The primary motivation of the proposed framework is the need for a video recommendation system for regional Indian languages. In particular, languages with cultural importance, such as Assamese in the Northeastern region of India, play a vital role and require stronger presence on the Web 3.0. As the web moves toward a multilingual and diversified semantic environment, it is necessary to ensure that such languages are properly represented. Alongside this, there is a growing requirement for an Assamese video recommendation framework that can focus on videos tagged and described in Assamese. A model that hybridizes semantic artificial intelligence with coordinative semantic reasoning is therefore both important and timely. This work addresses that gap and proposes such a framework as the need of the hour.

Contributions: The following are the primary contributions of the proposed model. First, keyword extraction is performed with informative terms taken directly from the dataset and then converted into English. This enables the regeneration of high-density auxiliary knowledge from the web in the form of open-ended data, which is classified using a strong autoencoder classifier. This aspect is considered novel. Second, the framework binds query words with entities from Latent Dirichlet Allocation and Wikidata, while applying a bilingual query formalization scheme through INLTK-based query processing. This is again novel and remarkable. Third, the framework achieves quantitative semantic intelligence through the use of SimRank and Kullback–Leibler divergence with differential thresholds and step deviance measures. An optimization process inspired by dolphin echolocation is applied for optimal solution computation, with SimRank as the criterion function. Finally, the proposed Assamese video recommendation framework outperforms baseline models and can be considered one of the best approaches. Moreover, this is one of the first Assamese video recommendation frameworks that encompasses semantic intelligence together with quantitative semantic reasoning, making it a novel contribution.

2 Related Work

Work on Assamese NLP has grown slowly but steadily, and the literature shows a clear shift from exploratory tools to deeper, more practical systems. One of the earliest attempts to formalize computational work on Assamese came from Barman et al. [1], who tested

CRF++ and fnTBL for part-of-speech tagging. Their results were useful mainly as proof-of-concept, since data scarcity limited accuracy. Soon after, Daimary et al. [2] explored hidden Markov models, which offered greater consistency but still struggled with the irregular grammar and free word order typical of Assamese. The real turning point came when neural methods began to be applied. Pathak et al. [3] proposed AsPOS, a BiLSTM-CRF tagger supported by multilingual embeddings such as MuRIL and Flair. This marked a break from earlier methods and showed that transfer learning from larger languages could benefit a low-resource one. Around the same time, Rahman and Sarma [4] produced a finite-state morphological analyzer through the Apertium framework. Though rule-based, their system captured many of the inflectional patterns that statistical models often ignored, making it valuable for translation and parsing tasks. Pathak et al. [5] later addressed named entity recognition, creating the AsNER dataset and providing baselines that enabled information extraction for the first time at scale. Sentiment analysis has attracted significant attention in the last few years. Das and Singh [6] introduced a multimodal framework that used both text and images, motivated by the noisy and informal style of Assamese social media. They followed this with a late-fusion design [7] that integrated captions and visual features with text, and in practice this approach yielded more reliable predictions. More recently, their empirical study [8] examined how specific words and lexical cues shape sentiment classification, a contribution that not only improved model building but also shed light on the expressive nature of Assamese. Resources have played a decisive role in sustaining this progress. Tamang and Bora [9] proposed a centralized repository that brought together datasets scattered across earlier projects. Their aim was not just convenience but also standardization, giving the community a single reference point for experimentation. Laskar et al. [10] added to this resource-building trend with EnAsCorp1.0, an English–Assamese parallel corpus released under the LoResMT initiative. This bilingual dataset became a foundation for translation tasks and for cross-lingual transfer more generally. Taken together, these studies illustrate how Assamese NLP has moved from modest beginnings to a field that, while still resource-constrained, now has the building blocks needed for larger applications. Statistical POS tagging has given way to deep neural architectures; isolated efforts have been supplemented by shared datasets; and sentiment analysis has grown from text-only classification to multimodal frameworks. For applications such as video recommendation, where tags, captions, and descriptions often appear in Assamese, these advances in linguistic processing provide the necessary groundwork. However, while these studies significantly advance Assamese NLP resources and linguistic modeling, they do not address the integration of semantic reasoning, bilingual query formalization, and generative metadata enrichment within a unified video recommendation framework. Existing video recommendation systems such as CVSV [15] and RTVE [17] are not designed for Assamese language processing, and models such as RDAT [16] focus primarily on text analytics without incorporating structured semantic intelligence or optimization-driven entity refinement. The proposed AssameseVidRec framework extends beyond these works by combining bilingual query processing, ontology-driven metadata generation, SimRank-based semantic reasoning, KL-divergence filtering, and metaheuristic optimization into a single coherent architecture tailored for Assamese content.

3 Methodology

The proposed Assamese video recommendation framework hybridizes generative artificial intelligence with semantic artificial intelligence. Figure 1 illustrates the system architecture. The video dataset used in this framework is designed such that the videos may not necessarily be in Assamese, but the associated tags and descriptions contain Assamese terms. At least a portion of the tags and labels are in Assamese. These Assamese tags and descriptions are pre-processed using customized Assamese language processing paradigms to yield keywords and informative terms. Tags and descriptions in English are also processed, and the extracted Assamese terms are strategically converted into English using the Google Translation API. The main reason for this translation is to reduce the cognitive gap between knowledge in Web 3.0 and the knowledge pipeline of the proposed system.

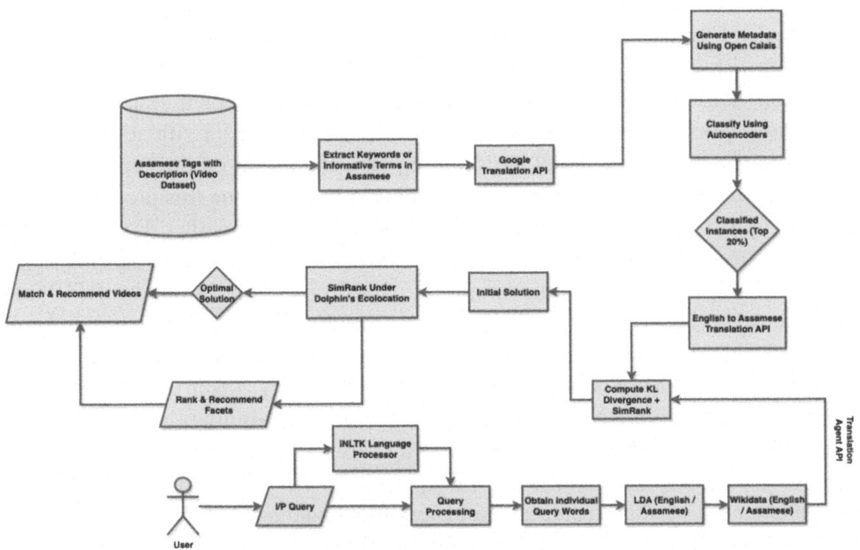

Fig. 1. Architecture of the proposed AssameseVidRec Framework

Metadata is generated using OpenCalais, where the terms extracted from the dataset in English serve as input. This produces high-density auxiliary knowledge, which is subjected to classification since raw metadata cannot be directly used. Autoencoders are employed for subject classification, chosen as a strong deep learning model capable of implicit feature extraction. Because of the large volume of classified instances, only the top 20% are retained and integrated into the pipeline. This selection helps prevent domain divergence and ensures that the metadata remains atomic and relevant. The selected metadata is then translated back into Assamese through the translation API, ensuring consistent use of Assamese in the system.

The framework supports bilingual queries. A user may enter queries in Assamese, English, or a mixture of both. All queries are processed using the iNLTK language processor, which detects the source language and performs tokenization, lemmatization, stop-word removal, and named entity recognition. English terms are then subjected to Latent Dirichlet Allocation (LDA) in English, while Assamese terms are subjected to LDA in Assamese. LDA, as a topic modeling framework, extracts topics or entities from the queries. The most relevant terms are passed forward for further processing.

The entities obtained from the LDA framework are next passed into Wikidata. Terms from English queries are linked to Wikidata in English, while Assamese terms are linked to Wikidata in Assamese. Wikidata, as a community-verified knowledge repository, enhances both the density and quality of the terms while lowering the cognitive gap between query words and Web 3.0 knowledge. The entities emerging from the LDA–Wikidata pipeline are then translated to Assamese, ensuring uniformity of language for subsequent computation.

Semantic reasoning is carried out through SimRank combined with Kullback–Leibler divergence. For this, thresholds are empirically fixed at 0.80 for SimRank and 0.10 for divergence to balance the number of entities with their relevance. These measures ensure that the entities selected remain both significant and contextually aligned. However, because the initial solution set is still large, the Dolphin Echolocation Algorithm is applied as a meta-heuristic optimization step. Inspired by the natural echolocation behavior of dolphins, this algorithm revisits candidate solutions iteratively until the optimal solution set is obtained.

The optimized solution set contains the best recommendation facets specific to the user query. These facets are ranked by SimRank values and used to match tags and descriptions of videos. The final output is a set of recommended Assamese videos aligned to the query context. In subsequent iterations, user clicks are fed back into the system. At this stage, only Assamese terms are processed through the LDA–Wikidata pipeline to maintain consistency between the user and the framework, reinforcing the recommendation of Assamese videos.

3.1 SimRank

SimRank, as shown in Eq. 1, is a graph-based similarity measure, was used in the framework to decide how close two entities are by checking the neighbors around them. The definition is that two objects are said to be similar if the objects linked to them are also similar, and this repeats in a recursive manner until stable scores are reached. A decay factor C, often taken between 0.6 and 0.8, is used so that the influence of far-off neighbors gets reduced.

$$\mathrm{s}(a,b) = \begin{cases} 1, a = b \\ \frac{\mathrm{C}}{|I(a)||I(b)|} \sum_{i \in I(a)} \sum_{j \in I(b)} s(i,j), \textit{otherwise} \end{cases} \tag{1}$$

In the above formulation, $I(a)$, $I(b)$ are in-neighbors of nodes a, b and $\mathrm{C} \in (0, 1)$ is a decay factor (often 0.6–0.8).

3.2 Kullback–Leibler Divergence (KL Divergence)

Kullback–Leibler divergence, as shown in Eq. 2, is a way to measure the distance between two probability distributions, was also applied. The idea here is that if two distributions are identical, the divergence is zero, but if they are different the value grows larger. In the system it is used to compare how entity distributions vary, and by doing this, entities that are not aligned to the context can be dropped while those that fit are retained.

$$D_{KL}(P \,||\, Q) = \sum_i P(i) \log \frac{P(i)}{Q(i)} \tag{2}$$

where P and Q are probability distributions of entity occurrences.

3.3 Dolphin Echolocation (Optimization Function)

For optimization, the dolphin echolocation algorithm was used. As shown in Eq. 3, it is a metaheuristic that comes from the way dolphins locate prey using sound waves, and it was adapted to update candidate solutions with sinusoidal functions. In practice, each solution is shifted at every iteration with an adaptive step, and the goal is to increase semantic closeness through SimRank while at the same time reducing redundancy through KL divergence. This keeps the final solution balanced, meaning both accurate and diverse.

$$\begin{aligned} x_{t+1} &= x_t + \alpha .\sin(2\pi f t + \phi) \\ &\max(SimRank(a, b) - \lambda D_{KL}(P||Q)) \end{aligned} \tag{3}$$

where α is the adaptive step, f is the frequency, ϕ is the random phase and λ is a balancing parameter.

4 Results and Performance Evaluation

Experiments are conducted on the four distinct datasets, namely the Dataful 2025 CBFC Certification Data for Movies in Assamese Language dataset [11], the GLM Human Activity Recognition (HAR – Video) Dataset [12], Chayanika Sarmah's 2023 Satriya-08 Double Handed Mudras dataset [13], and Rishav Sharma's 2024 YouTube Trending Videos dataset [14]. These videos were subjected to labeling through web crawling, and a translation-API based agent was applied using Google's Translation API for translating the video tags and descriptions from English to Assamese in the English datasets. If the descriptions were already in Assamese, they were retained as they were. iNLTK was used to process the linguistic variables and ensure that the entire dataset was converted into Assamese by the translation agents. Subsequently, a customized crawler was employed to gather additional videos from Web 3.0, ensuring that the descriptions and entities in video labels were converted into Assamese. The entities and terminologies thus crawled, along with their descriptions and annotations, were uniformly consolidated into a single large dataset purely in Assamese, which was then used for the experimentations.

The performance of the proposed strategic AssameseVidRec framework is evaluated using precision, recall, accuracy, F-measure, and false discovery rate (FDR) as the main metrics. Precision, recall, accuracy, and F-measure are chosen since they quantify the relevance of the results, while the false discovery rate captures the number of false positives included in the framework, thereby quantifying deviations in the model's data range.

Table 1. Comparison of Performance of the proposed AssameseVidRec with Other Approaches

Model	Average Precision %	Average Recall %	Average Accuracy %	Average F-Measure %	FDR
CVSV [15]	89.12	91.33	90.22	90.21	0.11
RDAT [16]	91.20	93.03	92.12	92.11	0.09
RTVE [17]	92.19	94.19	93.19	93.18	0.08
Proposed AssameseVidRec	**96.89**	**98.12**	**97.50**	**97.50**	**0.04**

From Table 1, it is observed that the proposed AssameseVidRec yields an overall average precision of 96.89%, an average recall of 98.12%, an average accuracy of 97.50%, and an average F-measure of 97.50%, with a lower FDR of 0.04. These results indicate that the proposed AssameseVidRec achieves the highest precision, recall, accuracy, and F-measure values while maintaining the lowest FDR compared to other baseline models.

The proposed framework has been baselined against three models, namely CVSV [15], RDAT [16], and RTVE [17]. Both CVSV [15] and RTVE [17] are existing video recommendation frameworks not designed for Assamese language support. RDAT [16], on the other hand, is an Assamese text analytical model that processes Assamese language text for natural language applications and has been adapted here to support recommendation. All three models were evaluated using the same dataset and environment as the proposed framework, with translation APIs used in CVSV and RTVE to align query words with auxiliary knowledge.

While RDAT [16] applies OCR with YOLO-darknet and gradient character distance for Assamese text extraction, it lacks contextual semantic reasoning, resulting in weaker performance compared to the proposed AssameseVidRec. Similarly, RTVE [17] relies on demographic filtering and candidate video selection, which produces strong recommendations in general contexts, but lacks auxiliary knowledge enrichment and semantic filtering for Assamese content. CVSV [15] performs the weakest, as it does not incorporate Assamese or bilingual support. In contrast, the proposed AssameseVidRec integrates bilingual query pre-processing, metadata generation with OpenCalais, autoencoder classification, domain divergence preservation, semantics-driven reasoning with SimRank and KL divergence, and optimization with Dolphin's Echolocation, resulting in its superior performance (Table 2).

As shown in Fig. 2, presents the line graph comparing precision percentages against the number of recommendations. It is evident from the graph that the proposed AssameseVidRec consistently outperforms all other baseline models, maintaining the highest precision across different recommendation counts. CVSV [15] remains at the lowest position throughout, RDAT [16] occupies the second position from the bottom, RTVE [17] ranks just below the proposed framework, and AssameseVidRec maintains the topmost position in the hierarchy.

The primary reason for this hierarchy is the design advantages of the proposed model. CVSV [15], RDAT [16], and RTVE [17] lack query pre-processing in two languages,

Table 2. Precision Percentage vs. No. of Recommendations

No. of Recommendations	CVSV [15]	RDAT [16]	RTVE [17]	Proposed AssameseVidRec	No. of Recommendations
10	91.87	93.52	94.29	98.87	**10**
20	90.22	92.90	93.48	97.89	**20**
30	89.19	91.78	92.19	96.87	**30**
40	88.79	90.44	91.91	95.78	**40**
50	87.81	89.89	90.83	94.36	**50**

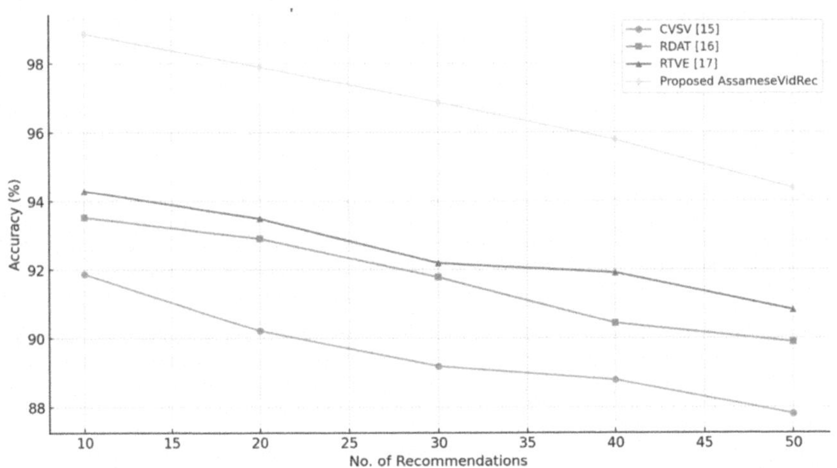

Fig. 2. Precision Percentage vs. No. of Recommendations

have weaker mechanisms for auxiliary knowledge enrichment, and miss semantic filtering of entities. In contrast, AssameseVidRec integrates bilingual query processing, Wikidata-assisted entity extraction, metadata generation, and semantics-driven quantitative reasoning with optimization. These aspects ensure that the system produces the most relevant recommendations, making it one of the best frameworks for Assamese video recommendation.

5 Conclusion

This paper proposes the Assamese Media Recommendation framework, which is a bilingual query processing technique capable of handling queries in both English and Assamese. The framework employs LDA and query word processing along with a knowledge-encompassment strategy using Wikidata. Metadata is generated using OpenCalais as the tool of choice, where terms in Assamese and English are extracted from the dataset, translated into English, and then integrated to bridge the cognitive gap between Web 3.0 knowledge and the proposed system pipeline. The model incorporates autoencoders for metadata classification to prevent domain divergence. This results in a domain-specific, knowledge-centric video recommendation model using SimRank with Kullback–Leibler divergence, thereby facilitating quantitative reasoning. A dolphin echolocation inspired metaheuristic model further helps in achieving optimal solutions for mapping and matching entities in video tags and descriptions, which in turn enables an annotation-driven recommendation process. Overall, the framework encompasses semantic artificial intelligence with quantitative reasoning in Assamese, contributing directly to the multilingual vision of Web 3.0. Experimental results show that the proposed Assamese video recommendation model yields an average precision of 96.89%, an average recall of 98.12%, an average accuracy of 97.50%, an average F-measure of 97.50%, and a low false discovery rate (FDR) of 0.04. These results establish it as a best-in-class solution. It is also the first framework designed specifically for Assamese that hybridizes autoencoders with incremental knowledge generation from both the dataset and query perspectives.

References

1. Barman, U., Sarmah, D., Sarma, K.K.: POS tagging of assamese language and performance analysis of CRF++ and fnTBL approaches. In: Proceedings of UKSim / ICACCI (2013)
2. Daimary, D., Sharma, U., Sarma, K.K.: Development of part of speech tagger for Assamese using hidden Markov model. Int. J. Comput. Appl. (2018)
3. Pathak, A., Nandi, P., Sarmah, D.: AsPOS: assamese part of speech tagger using deep learning approach. arXiv preprint arXiv:2212.07043 (2022)
4. Rahman, M., Sarma, K.K.: An implementation of apertium based assamese morphological analyzer. arXiv preprint arXiv:1503.03989 (2015)
5. Pathak, A., Nandi, P., Sarmah, D.: AsNER: annotated dataset and baseline for assamese named entity recognition. arXiv preprint arXiv:2207.03422 (2022)
6. Das, R., Singh, T.D.: A multi-stage multimodal framework for sentiment analysis of Assamese in a low-resource setting. OpenReview preprint (2022)
7. Das, R., Singh, T.D.: Image-text multimodal sentiment analysis framework of Assamese news articles using late fusion. In: Proceedings of IHCI (2023)
8. Das, R., Singh, T.D.: Which words are important?: An empirical study of Assamese sentiment analysis. Lang. Resour. Eval. (2024)
9. Tamang, S., Bora, A.: Enhancing Assamese NLP capabilities: introducing a centralized dataset repository. arXiv preprint arXiv:2410.11291 (2024)
10. Laskar, S.R., Ekbal, A., Saha, S.: EnAsCorp1.0: English–Assamese corpus for low-resource machine translation. In: Proceedings of LoResMT 2020. ACL Anthology (2020)
11. Dataful (Factly): CBFC Certification data for movies in Assamese language (2025). https://dataful.in/datasets/15388
12. Sharjeel, M.: Human Activity Recognition (HAR - Video Dataset) (2023). https://doi.org/10.34740/kaggle/dsv/5722068
13. Sarmah, C.: Sattriya-08 Double Handed Mudra's Dataset (2023). https://doi.org/10.17632/sjvrbxrvh8.2
14. Sharma, R.: YouTube Trending Video Dataset (updated daily) (2024). https://www.kaggle.com/rsrishav/youtube-trending-video-dataset
15. Deldjoo, Y., Elahi, M., Cremonesi, P., et al.: Content-based video recommendation system based on stylistic visual features. J. Data Semant. **5**, 99–113 (2016). https://doi.org/10.1007/s13740-016-0060-9

16. Enghi, M., Talukdar, A.K., Sarma, K.K.: Real-time detection of natural scene assamese texts using deep learning. In: Bhattacharjee, R., Neog, D.R., Mopuri, K.R., Vipparthi, S.K. (eds.) Artificial Intelligence and Data Science Based R&D Interventions, NERC 2022. Springer, Singapore (2023). https://doi.org/10.1007/978-981-99-2609-1_4
17. Huang, Y., Cui, B., Jiang, J., Hong, K., Zhang, W., Xie, Y.: Real-time video recommendation exploration. In: Proceedings of the 2016 International Conference on Management of Data (SIGMOD 2016), pp. 35–46. Association for Computing Machinery, New York (2016). https://doi.org/10.1145/2882903.2903743

A Comprehensive Survey of Traditional, Machine Learning, Deep Learning, and Federated Learning Methods for Multimodal Autism Spectrum Disorder Diagnosis

A. Sathish Kumar(✉), Malmathanraj Ramnathan, Avik Hati, and P. Palanisamy

National Institute of Technology, Tiruchirappalli, Tiruchirappalli, Tamil Nadu, India
{408124001,rmathan,avikhati,palan}@nitt.edu

Abstract. Autism spectrum disorder (ASD) is a complex neurodevelopmental condition with diverse behavioral, cognitive, and neurological characteristics, making reliable diagnosis a persistent challenge. This survey provides a structured review of four major methodological streams for ASD detection namely traditional clinical tools, machine learning (ML), deep learning (DL), and federated learning (FL). Special attention is given to multimodal data integration across behavioral observations, eye tracking, speech, EEG, MRI, genetics, and electronic health records. The study compares algorithms, fusion strategies, and evaluation practices, while highlighting key challenges such as dataset scarcity, cultural bias, interpretability, and cross site generalization. Deep learning methods, including CNNs, RNNs, and graph based models, have shown promise in handling complex multimodal inputs, though transparency and scalability remain open issues. Federated learning addresses data privacy and governance concerns, enabling collaborative model training across institutions. This survey article enumerates the Multimodal ASD Diagnosis and gives a clear picture for further research, with the ultimate goal of building accurate, equitable, and clinically deployable diagnostic frameworks.

Keywords: Autism Spectrum Disorder · Multimodal · Machine Learning · Deep Learning · Federated Learning

1 Introduction

A neurodevelopmental and neurodegenerative condition that affects people for the rest of their lives, ASD interferes with social interaction, communication, and behavioural characteristics. Despite the fact that ASD can impact individuals at any age, symptoms typically manifest by the time a child is two years old, and boys are diagnosed with the disorder more frequently than girls [1]. There may be a number of contributing elements, such as environment, brain differences, and inheritance [2, 3]. While treatment

© The Author(s), under exclusive license to Springer Nature Switzerland AG 2026
A. Kannan et al. (Eds.): ADCOM 2025, CCIS 2947, pp. 183–197, 2026.
https://doi.org/10.1007/978-3-032-26269-1_13

is available, early intervention can improve quality of life [4]. Computational methods are being explored for faster and more accurate identification because behavioural tests such as Autism Diagnostic Observation Schedule (ADOS) and Childhood Autism Rating Scale (CARS) are required for diagnosis, but are time-consuming.

ASD is still challenging to diagnose due to its variable presentation and reliance on subjective clinical evaluations [5]. The full spectrum of ASD characteristics is often missed by conventional diagnostic methods, leading to inaccurate or delayed diagnosis. A promising solution is provided by multimodal techniques, which integrate many data sources such as eye tracking, functional magnetic resonance imaging (fMRI), electroencephalography (EEG), and behavioural evaluations [6, 7]. The EEG offers information about patterns of cerebral activity that might be suggestive of ASD, and facial expression analysis and eye tracking provide behavioural and emotional clues. The fMRI enhances the diagnostic paradigm by providing temporal and spatial information regarding brain function. The efficient merging of various modalities is made possible by sophisticated machine learning models such as hierarchical feature extractors, stacked denoising autoencoders (SDAEs), and graph convolutional networks (GCNs), which improve diagnostic resilience and accuracy [3, 7]. Additionally, federated learning frameworks enable scalable deployment while maintaining patient privacy by facilitating decentralised model training across institutions. In order to improve patient outcomes and care methods, doctors can obtain earlier, more accurate, and personalised diagnoses of ASD by utilising multimodal data and state-of-the-art computational algorithms. Deep learning models, including CNNs and graph-based networks, have shown accuracy of 72.08% on Automated Anatomical Labeling template and obtains accuracy of 75.07% on Consensus Clustering 200 template respectively in extracting features from multi-view brain transformer [9]. These models can detect subtle patterns that may be missed in clinical evaluations. Federated learning further enhances this by enabling collaborative model training across institutions without compromising patient privacy [10]. The objective of this survey is to compare these approaches, highlight their strengths and limitations, and emphasize the value of integrating multiple data types such as EEG, eye-tracking, and clinical scores for more accurate and early ASD detection [11]. This review also identifies research gaps and suggests future directions for building scalable and interpretable diagnostic systems.

The rest of the paper is organized as follows. Section 2 shows the methodology of review. Section 3 gives a detailed review on ASD and Multimodal diagnosis highlighting the importance of multimodal data in ASD diagnosis. Section 4 outlines traditional methods for ASD diagnosis. Sections 5, 6, 7 provide a comprehensive review on the relevant literature related to ML, DL and FL techniques for ASD diagnosis. Section 8 concludes the review.

2 Methodology of Review

This review was conducted following the Preferred Reporting Items for Systematic Reviews and Meta-Analyses (PRISMA) guidelines [34] to ensure a structured and transparent study selection process. The methodology included systematic database searching, predefined inclusion and exclusion criteria, and multi-stage screening to identify relevant studies related to multimodal ASD diagnosis.

2.1 Search Strategy

A comprehensive literature search was performed across major academic databases, including IEEE Xplore, Google Scholar, Science Direct, PubMed, MDPI, Scopus, Springer, and Wiley. Keywords such as "Autism Spectrum Disorder", "ASD diagnosis", "multimodal", "machine learning", "deep learning", "federated learning", "EEG", "fMRI", and "eye-tracking" were used in various combinations.

The search focused on peer-reviewed journal articles and conference papers published in recent years, with particular emphasis on studies applying computational methods for ASD detection. Duplicate records were removed before proceeding to the screening stage.

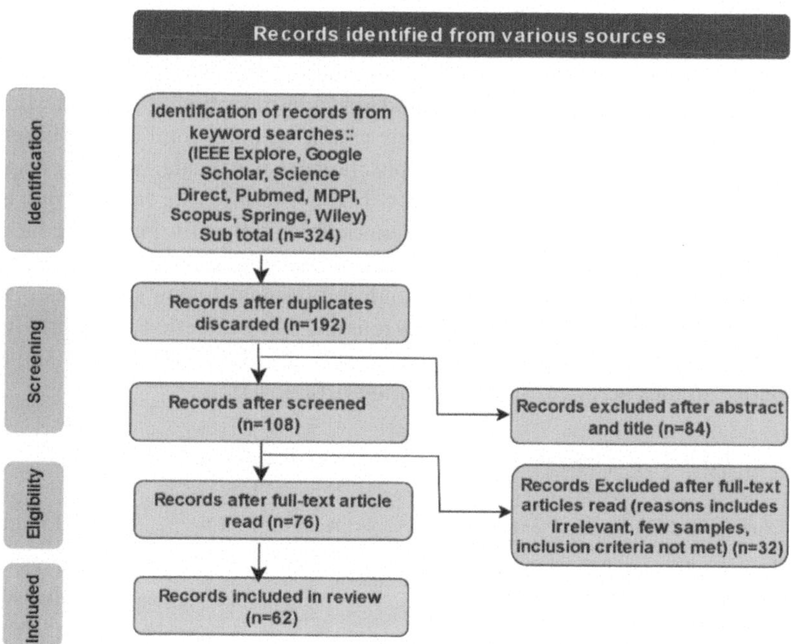

Fig. 1. PRISMA flow diagram of this study.

2.2 Study Selection Process

The study selection process followed a multi-stage screening procedure, as illustrated in Fig. 1. Initially, 324 records were identified from various databases. After removing duplicates, 192 records remained for title and abstract screening. Studies that did not meet the inclusion criteria were excluded at this stage.

Subsequently, 76 full-text articles were assessed for eligibility. Articles that were irrelevant, had insufficient data, or did not satisfy the inclusion criteria were excluded. Finally, 62 studies were included in the qualitative synthesis of this review.

3 ASD and Multimodal Diagnostics

3.1 Multimodal Data Sources in ASD

Autism Spectrum Disorder is a condition that affects people in ways and using only one method to assess ASD often does not work very well. Using different methods like looking at behavior, how the brain works, pictures of the brain and medical information helps us understand ASD [7]. Video analysis helps us automatically look at gestures, facial expressions and behaviors that people do over and over for an early check of autism [12]. Video based analysis of autism looks at how people talk and the words they use including how they sound and if they have trouble talking and Video based analysis uses computer programs to find these things [5, 6].

Neuroimaging modalities such as EEG and rs-fMRI provide insights into atypical brain connectivity and neural oscillations [3, 4]. Deep learning and graph-based models have demonstrated improved classification performance using these modalities. Several studies have further shown that combining EEG, eye-tracking, and imaging data enhances diagnostic robustness compared to single-modality models. Questionnaire based assessments support ASD screening by capturing behavioral and developmental traits through structured items such as multiple choice, Likert scale, or binary responses, available in both paper and electronic formats. These responses are encoded into numerical features using score aggregation or categorical encoding for predictive modeling. When combined with modalities such as EEG or imaging, questionnaire-derived features enhance diagnostic performance through multimodal fusion strategies. [28–33]. The integration of heterogeneous modalities remains a key direction for improving reliability and generalizability of ASD diagnostic systems. Tables 1, 2 and 3 summarize representative studies across different modalities (Table 4).

Table 1. Representative EEG and Eye-Tracking Based ASD Studies

Author (Year)	Modality	Model	Accuracy (%)	Key Limitation
Han et al. (2022) [5]	EEG + ET	Multimodal SDAE	95.56	Requires simultaneous multimodal acquisition
Zhang et al. (2023) [13]	ET	UASN	100 (subject-level)	Limited ecological validity of stimuli

(continued)

Table 1. (*continued*)

Author (Year)	Modality	Model	Accuracy (%)	Key Limitation
Noor et al. (2024) [14]	EEG	TL + SqueezeNet	85.5	Possible feature loss during transfer
Hu et al. (2023) [15]	EEG	RAGNN	92.78	Binary classification only
Li et al. (2025) [16]	EEG	LSTA-CNN	98.9	Risk of overfitting due to attention complexity
Chen et al. (2022) [17]	ET	LSTM + DVP	96.73	Limited to early childhood screening
Tseng et al. (2024) [11]	EEG	SVM	95.8	Small sample size
Xia et al. (2024) [18]	Facial + ET	ViT + LongFormer	96.86	Limited accessibility in remote regions

Table 2. Representative MRI-based ASD Classification Studies.

Author (Year)	Dataset/Sample Size	Model	Accuracy (%)	Key Limitation
Song et al. (2025) [3]	ABIDE (871)	HE-MF	95.17	Modality similarity bias in graph learning
Sadiq et al. (2022) [19]	rs-fMRI (108)	SVM/KNN/Ensemble	98.6	Small per-class sample size
Wang et al. (2023) [7]	ABIDE (871)	WL-DGCNN	77.27	Performance below clinical threshold
Hao et al. (2024) [20]	rs-fMRI (1035)	MC-SDL + Fusion	78.82	Single-modality limitation
Lu et al. (2022) [21]	rs-fMRI (391)	DeepTSK	68.6	Functional connectivity instability
Du et al. (2024) [22]	Multi-site (1112)	HBFN-MLD	68.98	Inter-site variability
Jung et al. (2023) [23]	rs-fMRI (880)	ROI Selection Model	73.71	Excludes multimodal features

(*continued*)

Table 2. (*continued*)

Author (Year)	Dataset/Sample Size	Model	Accuracy (%)	Key Limitation
Liu et al. (2023) [24]	rs-fMRI (1102)	LRCDR	73.1	Site heterogeneity challenges
Tang et al. (2019) [25]	ABIDE (871)	LDA	77.7	Limited discriminative capacity
Wang et al. (2022) [26]	rs-fMRI (613)	Contrastive MV-GCN	75.2	Overfitting risk

Table 3. Questionnaire and Multimodal Clinical-Based ASD Detection Studies.

Author (Year)	Modality/Tool	Model	Accuracy (%)	Key Limitation
Hajjej et al. (2024) [29]	Q-CHAT-10	RF + XGBoost	99.99	Dataset imbalance
Akter et al. (2019) [30]	Q-CHAT-10, AQ-10	ML Classifiers	93.89–98.77	Limited ASD samples
Sellamuthu et al. (2024) [6]	Facial + ADOS	Multimodal Fusion	97.05	Fusion complexity
Alutaibi et al. (2025) [31]	Q-CHAT-10	Capsule DenseNet++	99.2	Limited demographic diversity
Tang et al. (2020) [32]	M-CHAT + Voice	SVM	96.39	Limited binary decision modeling

Table 4. List of Abbreviations.

Abbreviation	Full Form
TD	Typically Developing
EEG	Electroencephalography
ET	Eye Tracking
rs-fMRI	Resting-State Functional MRI
RNN	Recurrent Neural Network
SDAE	Stacked Denoising Autoencoder
UASN	Unified Attention Scanpath Network
RAGNN	Regional-Asymmetric Adaptive Graph Neural Network
LSTA-CNN	Lightweight Spatio-Temporal Attention CNN

(*continued*)

Table 4. (*continued*)

Abbreviation	Full Form
HE-MF	Hierarchical Extraction and Multimodal Fusion
WL-DGCNN	Weight Learning Deep Graph CNN
LRCDR	Low-Rank Class Discriminative Representation
Q-CHAT	Quantitative Checklist for Autism in Toddlers
M-CHAT	Modified Checklist for Autism in Toddlers
AQ	Autism Quotient
ASQ	Ages and Stages Questionnaire
CSBS	Communication and Symbolic Behavior Scales
PEDS	Parents' Evaluation of Developmental Status

4 Traditional Methods for ASD Diagnosis

4.1 Clinical and Behavioral Assessments

The Autism Spectrum Disorder is mostly found out through doctor visits that look at the person development history how they behave and use special tests. Doctors often use the Autism Diagnostic Observation Schedule and the Autism Diagnostic Interview to figure out if someone has ASD because these tests have been proven to work. When doctors want to check young kids they use the Modified Checklist for Autism, in Toddlers, which is a simple test that doctors can do in their office. Doctors and nurses use observations in homes and schools to see how people behave in these places. They also use a way of coding to look really closely at how people interact with each other. This helps them find problems early and keep track of how people are doing when they get treatment. They do this with observations and micro-coding approaches to understand behaviors and interaction patterns, in home and school environments [35]. Figure 2 illustrates the master block diagram for ASD detection highlighting the integration of behavioral, neuroimaging, and clinical modalities.

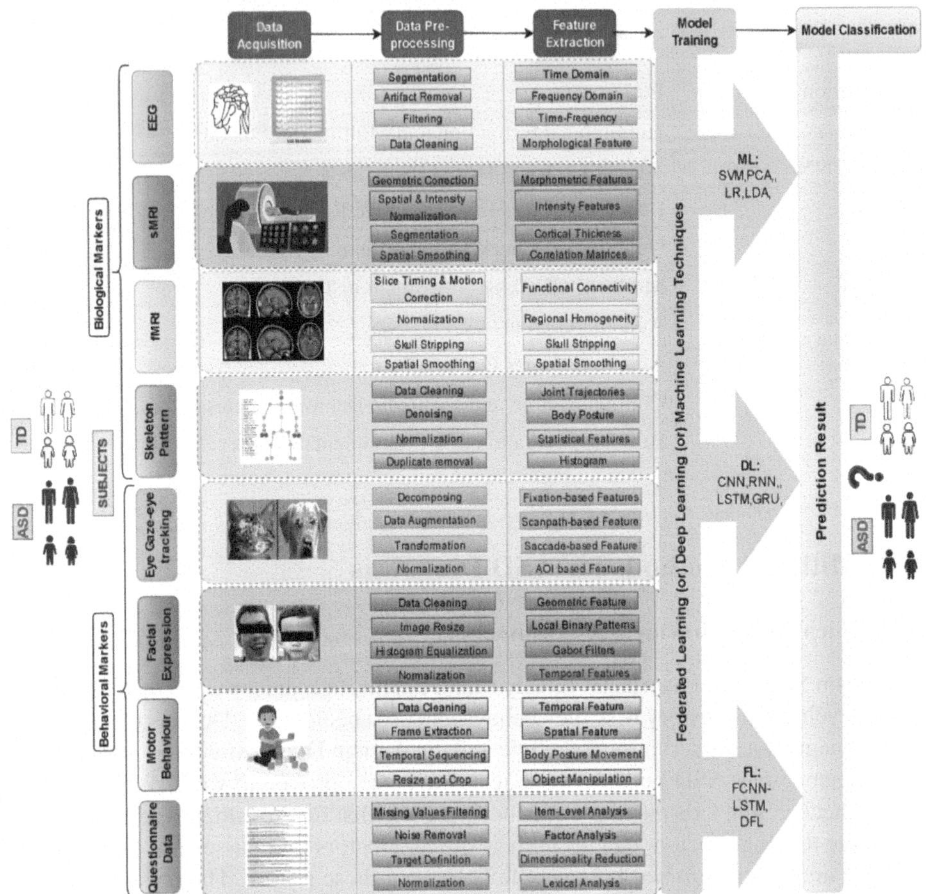

Fig. 2. Master block diagram for ASD detection with various modalities.

4.2 Strengths and Limitations

Traditional practices have a lot of things going for them. They give us a picture of symptoms in many different situations. This helps us avoid getting the diagnosis wrong by using methods. We also use tools that we know work well like ADOS and ADI-R. However, there are some problems with traditional practices. Doing an assessment can take a lot of time and resources sometimes taking weeks. This means that we have to wait a time to start helping someone. The people who do the assessments need training and some clinicians are better at it than others. This can make the assessments inconsistent. Traditional practices also do not work well for people, from cultures or who speak different languages. This can lead to mistakes when we are trying to diagnose someone because our tools might be biased.

5 Machine Learning Techniques for ASD Diagnosis

5.1 Overview of ML in ASD Diagnosis

Machine learning has become a part of finding ASD automatically. It does this by looking at lots of kinds of data like the signals from our brains, pictures of our brains questions about how we behave and how we move. Machine learning is better than the ways of doing things because it can look at the data and say what it means without being biased. It can even help find ASD on by recognizing patterns in big sets of data. Machine learning is really good at looking at lots of information and figuring out what it means which is very useful, for ASD [2, 4].

5.2 Common Algorithms

There are ML algorithms that people use to classify ASD and each one is good for different types of data and tasks.

Support Vector Machine (SVM): This is a choice for classifying ASD because it works well with a lot of information. It creates a way to separate ASD and typical development classes often using helpers like RBF for data that is not straightforward. For example, SVM was able to classify ASD from brain signals with features like fuzzy entropy and DFA exponents 93.59% of the time. In a version called Quantum SVM it was able to correctly classify ASD 94.7% of the time using a special technique on brain signals. SVM also worked well for analyzing movement features. Was able to correctly detect ASD 88.37% of the time [36].

Random Forest (RF): This is a method that uses decision trees to make predictions and it is good at handling data that is not balanced. It also tells us which features are the important. RF was able to screen for ASD 96% of the time by looking at how people played serious games and it was able to correctly detect ASD 100% of the time when looking at people of different ages. In studies that used brain signals RF was able to classify what people were thinking about when they moved their bodies after using a special filter [37]. Other algorithms that people use include Decision Trees, which create models that are easy to understand like when it correctly screened for ASD in children 100% of the time. There is also K-Nearest Neighbors, which looks at how similar things are to each other like when it correctly classified ASD 88.37% of the time using movement features. Then, there are Neural Networks and Quantum Neural Networks, which are good at recognizing complicated patterns, like when they correctly classified ASD 98.9% of the time using special features. These algorithms are often better, than methods because they can use many different types of data. ASD classification is getting better with these machine learning algorithms [4, 36].

5.3 Performance and Challenges

Machine Learning models for ASD detection show promising performance but face challenges. Accuracies range from 88% (kinematic features with KNN) to 100% (SVM/RF). QSVM and QNN achieved 98.9% on EEG data with quantum enhancements. Ensemble methods like RF yielded 96% in serious games, while SVM excelled in multi class tasks (99.55%) [1, 36]. Challenges include small sample sizes limiting generalizability, dataset biases, and ethical concerns in privacy. Quantum models require specialized hardware, and interpretability remains an issue. Future work should focus on larger, diverse datasets and hybrid approaches for robust, explainable diagnostics [36].

6 Deep Learning Methods for ASD Diagnosis

6.1 Convolutional Neural Networks (CNNs) for Imaging and Video

CNNs are really good at helping us understand ASD because they can look at pictures and videos. For example, Ahmed and his team made a CNN system that can make one picture from lots of brain scan pictures and it works better than systems. Something similar happened with Capsule DenseNet++ it is also a type of CNN system that's really good at finding ASD it can even find it more, than 99% of the time. We can also use CNNs to look at videos and find things that might mean someone has ASD like faces that do not look right or movements that they do over and over [31]. These things can help us know if someone has ASD.

6.2 Recurrent Neural Networks (RNNs) and Transformers for Speech and Text

Recurrent Neural Networks and their advanced form Long Short Term Memories are really good at understanding things that happen one after the other like speech, language and EEG signals. Li and his team used Long Short Term Memories with attention to see how parents and kids interact with each other when they play and they got it right 90% of the time [39]. Recurrent Neural Networks are also used in ways. For example, some people combined Convolutional Neural Networks with Recurrent Neural Networks to look at EEG signals and find brain activity that's specific to ASD. Then there are Transformers, which are not used much to study ASD but they are getting popular because they can understand how things are connected over long periods of time like, in speech and text data.

6.3 Graph Neural Networks (GNNs) for Brain Connectivity Analysis

Brain connectivity studies often look at brain regions as nodes and connections as edges. This makes Graph Neural Networks or GNNs a fit. Lu et al. used a learning method that combines fuzzy systems with deep learning models. This helped them understand patterns in fMRI data better [21]. Their approach worked well giving accuracy in classification and more understandable results. GNN models show promise in ASD research because they can learn from brain connectivity data that is already in a graph format. GNNs are suitable, for brain connectivity studies. They help in understanding ASD through graph data.

6.4 Multimodal Deep Learning Fusion Strategies

Early Fusion: This method combines features from sources before training a model. For example, brain wave features and behavioral features are put together to make classification more accurate. Late Fusion: Here separate models are trained for each source. Then their results are combined. A study by Li et al. used this approach to combine movement and interaction patterns in a game based framework. Hybrid Fusion: Some models use attention and self-learning to weigh features effectively. For instance, Capsule DenseNet++ used self-attention with CNNs to make ASD detection more robust [31]. Learning Shared Representations: Some researchers create embeddings from different data types. Lu et al. used a learning method with fuzzy systems to create embeddings, from brain scan data, which improved accuracy and kept the results easy to understand.

6.5 Interpretability and Explainability in DL for ASD

Explainable AI (XAI) is becoming more popular because it helps make predictions clear and transparent. This is especially important in fields like medicine, where doctors need to trust the predictions. Clinicians want to know how the model arrived at its conclusions. So XAI is crucial for getting doctors to use and rely on deep learning models. Models like Capsule DenseNet++ and hybrid deep models are helping to make deep learning more transparent [31]. Deep learning models need to be transparent to gain clinician trust. Clinician trust is essential for the adoption of deep learning models.

6.6 Strengths, Weaknesses, and Clinical Utility

Deep learning methods are really good at handling data that comes in different forms. They can automatically pick out features and do better than traditional machine learning in terms of accuracy when classifying things. However, DL models have some downsides. They need a lot of labeled data to work well which can be a challenge. They also require a lot of computing power. Another issue is that they can be sensitive to differences in data collected from places. In a setting these models could help doctors diagnose patients faster and more accurately. There are some barriers to using them in the real world. One big issue is that they lack interpretability, which means it's hard to understand why they make the predictions they do [21]. There are also challenges related to sharing data. To make deep learning models more useful in settings future research should focus on developing methods that are more explainable.

7 Federated Learning for ASD Diagnosis

7.1 Motivation Privacy, Security, and Data Governance

ASD research needs information like MRI images, EEG signals, facial features and behavioral questionnaires. When we share this kind of data across clinics or research institutions it can lead to privacy issues and problems with rules, like GDPR and HIPAA. Federated Learning helps here by keeping the data local and only sharing model details with a distributed hub. This approach stops identities from leaking and reduces the risk of unauthorized access [40]. Also FL makes it possible for many institutions to work together without needing one central dataset. This improves how we handle data and makes sure we follow guidelines. It helps ASD research. Keeps patient data safe.

7.2 Federated Learning Frameworks

Federated Learning can follow two structures. In Federated Learning many hospitals or labs work as clients and they send updates of their models to a global server. This global server puts all the updates together like in the case of CNN-LSTM frameworks that are used to detect ASD [33]. In decentralized FL, which is also called peer-to-peer Federated Learning there is no server that puts everything together. Instead the clients, which are the hospitals or labs exchange updates of their models with each other. This makes the system better able to handle problems like when a server stops working. It is also good for big healthcare projects that involve many people. Both of these structures have been used to detect ASD. The centralized structure is simple while the decentralized structure is better, at handling problems and making sure everything is fair when the models are being trained. Federated Learning is used in these structures to make sure the system works well.

7.3 Key Algorithms for FL in Healthcare

The FedAvg algorithm is a method in Federated Learning for healthcare. It works like this: each client trains a model on its data and then the updates are combined to create a global model. Research has shown that FedAvg can be quite accurate even when the data is spread across sites for ASD. FedAvg and its variants such as weighting and momentum based updates are used to improve convergence in ASD datasets that vary a lot from site to site. These variants of FedAvg have been applied to improve results, in ASD diagnosis. The FedAvg algorithm and its variants are widely used for ASD diagnosis [41]. People with ASD have different symptoms. So a single approach may not work for everyone. Some methods like FedPer and pFedMe try to make a model that works for a hospital or research site. These methods make a model that's a little bit different for each place. This way each place gets a model that is tailored to their needs. At the time they can still work together with other places. This is really helpful when the data from places is different. For example, when the data is, from people of ages or backgrounds. ASD models can be made to work for each place. FedPer and pFedMe are used to make ASD models that are tailored to each place [33]. In ASD research the information they collect is usually not the same. For instance one hospital might have a lot of MRI data while another hospital has a lot of questionnaires. Some methods like FedProx and hypergraph networks in machine learning help make the training more stable when the data is not the same. This is really important for ASD diagnosis because it helps make sure the diagnosis is fair and accurate for people from backgrounds and populations, with ASD [41].

7.4 Privacy-Preserving Mechanisms in FL

Differential privacy is a way to keep people's information safe. It does this by adding some noise to the updates that are made to a model. This means that even if someone tries to reverse engineer the updates they will not be able to get any details. This has been used in a type of learning called Federated Learning for predicting ASD. It makes sure that each person's information is protected [41]. It still allows for accurate diagnoses to be made. Homomorphic encryption is another way to keep information safe. It lets people do calculations on data that is encrypted without having to decrypt it. This is useful for detecting ASD because it means that even the server that is collecting all the data cannot see the individual updates. This protects against people who might be working on the server and try to look at the data. There is also something called aggregation. This means that the server can only see the combined updates, from all the places not the individual updates. This is important because it means that no one can take the updates and figure out what any one hospital has contributed. This has been shown to work in systems that use different types of data to predict ASD [33].

7.5 Applications of FL in Multimodal ASD Diagnosis

ASD diagnosis benefits from integrating multiple modalities such that behavioral traits, facial features, MRI, EEG, and caregiver reports. Federated Learning frameworks have successfully merged such data without requiring centralized storage. Combining behavioral and facial traits within FL achieved over 70% accuracy [10], while hypergraph FL approaches fused fMRI with clinical scores to capture complex relationships. Other works have explored IoT-based ASD monitoring integrated with FL for real-time adaptive diagnosis.

7.6 Challenges in Communication, Scalability, and Bias

Despite progress, FL for ASD diagnosis faces several hurdles. Communication overhead between local clients and servers slows training, especially with high-dimensional neuroimaging data [33]. Scalability remains a concern when many institutions participate, as synchronization becomes difficult. Bias is another issue in data imbalance across sites may lead to models favoring certain populations, reducing fairness. Future research suggests combining FL with gradient boosting and explainable AI to address these challenges, making ASD diagnosis more reliable and equitable.

8 Conclusion and Future Directions

This survey looks at ways to diagnose Autism Spectrum Disorder, including traditional methods, machine learning, deep learning and federated learning. It puts all these methods together to get a view of how they work with different kinds of information like pictures, EEG, behavior and speech. This helps us see what methods are getting better and how they can be used in life. Traditional tools like ADOS-2 and ADI-R are still good. They take a lot of time and money and they might not work well for people from different cultures. Machine learning and deep learning are faster and more accurate especially when using things like CNNs, RNNs and multimodal fusion models to look at behavior, pictures and brain activity. However, Autism Spectrum Disorder diagnosis using machine learning and deep learning has some problems, such as not having data not working well across different sites and being hard to understand. Federated learning is a way for different institutions to work together without sharing information but it has its own problems like dealing with different kinds of data and communicating effectively.

In the future researchers should work on getting different kinds of data over time making artificial intelligence that doctors can trust and finding better ways to combine different kinds of information. Autism Spectrum Disorder diagnosis using learning and working closely with doctors will be very important, for making diagnosis tools that are fair and work well for everyone and that can help diagnose ASD earlier and more accurately.

References

1. Shinde, A.V., Patil, D.D.: Multi-classifier recommender system for early ASD detection. Healthc. Anal. **4** (2023). https://doi.org/10.1016/j.health.2023.100211
2. Kavadi, D.P., et al.: Hybrid ML model for autism diagnosis. IEEE Access **12**, 194911–194921 (2024). https://doi.org/10.1109/ACCESS.2024.3520009
3. Gao, J., Song, S.: Hierarchical feature extraction and multimodal integration for ASD identification. IEEE J. Biomed. Health Inform. **29**, 4920–4931 (2025). https://doi.org/10.1109/JBHI.2025.3540894
4. Hasan, S.M., et al.: ML framework for early-stage ASD detection. IEEE Access **11**, 15038–15057 (2023). https://doi.org/10.1109/ACCESS.2022.3232490
5. Han, J., et al.: Multimodal identification of ASD in children. IEEE Trans. Neural Syst. Rehabil. Eng. **30**, 2003–2011 (2022). https://doi.org/10.1109/TNSRE.2022.3192431
6. Sellamuthu, S., Rose, S.: Multimodal approach for autism prediction. IEEE Access **12**, 121688–121699 (2024). https://doi.org/10.1109/ACCESS.2024.3453440
7. Wang, M., et al.: Multimodal ASD diagnosis via DeepGCN. IEEE Trans. Neural Syst. Rehabil. Eng. **31**, 3664–3674 (2023). https://doi.org/10.1109/TNSRE.2023.3314516
8. Lai, M., et al.: ML-based retinal image analysis for ASD screening. EClinicalMedicine **28** (2020). https://doi.org/10.1016/j.eclinm.2020.100588
9. Dong, Q., et al.: Multiview brain network transformer for ASD diagnosis. IEEE J. Biomed. Health Inform. **28**, 4854–4865 (2024). https://doi.org/10.1109/JBHI.2024.3396457
10. Shamseddine, H., et al.: Federated learning for multi-aspect ASD detection. arXiv (2022). https://doi.org/10.48550/arXiv.2211.00643
11. Tseng, Y.L., et al.: Fusion of cortical activation and connectivity using mobile EEG. IEEE Trans. Neural Syst. Rehabil. Eng. **32**, 3026–3035 (2024). https://doi.org/10.1109/TNSRE.2024.3417210
12. Mehralizadeh, B., et al.: Sensorized toy car for autism screening. Sustainability **15** (2023). https://doi.org/10.3390/su15107790
13. Zhang, Y., et al.: Uncertainty-inspired ASD screening. LNCS, pp. 399–408 (2023). https://doi.org/10.1007/978-3-031-43904-9_39
14. Al-Qazzaz, N.K., et al.: Transfer learning CNN for EEG-based ASD classification. IEEE Access **12**, 64510–64530 (2024). https://doi.org/10.1109/ACCESS.2024.3396869
15. Hu, W., et al.: Regional-asymmetric adaptive GCN for ASD diagnosis. IEEE Trans. Neural Syst. Rehabil. Eng. **32**, 200–211 (2024). https://doi.org/10.1109/TNSRE.2023.3347134
16. Li, J., et al.: LSTA-CNN for EEG-based ASD diagnosis. IEEE Trans. Neural Syst. Rehabil. Eng. **33**, 2456–2465 (2025). https://doi.org/10.1109/TNSRE.2025.3580593
17. Xia, C., et al.: Dynamic viewing pattern analysis for ASD screening. IEEE Trans. Biomed. Eng. **70**, 1622–1633 (2023). https://doi.org/10.1109/TBME.2022.3223736

18. Xia, C., et al.: Transformer-based facial cue representation for ASD. IEEE Trans. Affect. Comput. **16**, 83–97 (2025). https://doi.org/10.1109/TAFFC.2024.3412032
19. Sadiq, A., et al.: Non-oscillatory connectivity for ASD subtype classification. IEEE Access **10**, 14049–14061 (2022). https://doi.org/10.1109/ACCESS.2022.3146719
20. Hao, X., et al.: Multi-cluster diverse learning for ASD identification. IEEE Trans. Emerg. Top. Comput. Intell. **8**, 3860–3873 (2024). https://doi.org/10.1109/TETCI.2024.3377551
21. Lu, Z., et al.: Composite feature learning with fuzzy systems for ASD classification. IEEE/ACM Trans. Comput. Biol. Bioinform. **20**, 476–488 (2023). https://doi.org/10.1109/TCBB.2022.3163140
22. Du, J., et al.: Brain network distance for ASD severity identification. IEEE Trans. Neural Syst. Rehabil. Eng. **33**, 162–174 (2025). https://doi.org/10.1109/TNSRE.2024.3516216
23. Jung, W., et al.: Explainability-guided ROI selection for ASD diagnosis. IEEE Trans. Med. Imaging **43**, 1400–1411 (2024). https://doi.org/10.1109/TMI.2023.3337362
24. Liu, X., et al.: Low-rank representation for multi-site ASD identification. IEEE Trans. Neural Syst. Rehabil. Eng. **31**, 806–817 (2023). https://doi.org/10.1109/TNSRE.2022.3233656
25. Mostafa, S., Tang, L., Wu, F.X.: Brain network eigenvalue-based ASD diagnosis. IEEE Access **7**, 128474–128486 (2019). https://doi.org/10.1109/ACCESS.2019.2940198
26. Zhu, H., et al.: Contrastive multi-view GCN for ASD classification. IEEE Trans. Biomed. Eng. **70**, 1943–1954 (2023). https://doi.org/10.1109/TBME.2022.3232104
27. Tamizhmalar, D., et al.: Multimodal deep learning framework for ASD diagnosis. In: ICPECTS (2024). https://doi.org/10.1109/ICPECTS62210.2024.10779999
28. Almadhor, A., et al.: Explainable and secure multimodal ASD prediction framework. Complex Intell. Syst. **11** (2025). https://doi.org/10.1007/s40747-025-01790-3
29. Hajjej, F., et al.: Ensemble learning framework for ASD identification. IEEE Access **12**, 35448–35461 (2024). https://doi.org/10.1109/ACCESS.2024.3349988
30. Akter, T., et al.: ML models for early ASD detection. IEEE Access **7**, 166509–166527 (2019). https://doi.org/10.1109/ACCESS.2019.2952609
31. Alutaibi, A.I., et al.: Capsule DenseNet++ for autism detection. Comput. Biol. Med. 190 (2025). https://doi.org/10.1016/j.compbiomed.2025.110038
32. Tang, C., et al.: Video-audio ASD identification under still-face paradigm. IEEE Trans. Neural Syst. Rehabil. Eng. **28**, 2401–2410 (2020). https://doi.org/10.1109/TNSRE.2020.3027756
33. Lakhan, A., et al.: Federated CNN-LSTM framework for ASD detection. Comput. Biol. Med. **166** (2023). https://doi.org/10.1016/j.compbiomed.2023.107539
34. Minissi, M.E., Chicchi Giglioli, I.A., Mantovani, F., et al.: Assessment of the autism spectrum disorder based on machine learning and social visual attention: a systematic review. J. Autism Dev. Disord. **52**, 2187–2202 (2022). https://doi.org/10.1007/s10803-021-05106-5
35. Fusar-Poli, L., et al.: Diagnosing ASD in adults using ADOS-2 and ADI-R. J. Autism Dev. Disord. **47**, 3370–3379 (2017). https://doi.org/10.1007/s10803-017-3258-2
36. Bawa, P., et al.: Multiclass ASD classification using ML. e-Prime **8** (2024). https://doi.org/10.1016/j.prime.2024.100602
37. Rashed, A.E., et al.: Efficient ML models for ASD diagnosis. Biomed. Signal Process. Control **100** (2025). https://doi.org/10.1016/j.bspc.2024.106949
38. Saranya, S., Menaka, R.: Explainable ML network for EEG-based ASD classification. IEEE Access **13**, 32016–32030 (2025). https://doi.org/10.1109/ACCESS.2025.3542390
39. Li, X., et al.: Attention-enhanced deep learning for early ASD detection. Eng. Appl. Artif. Intell. **148** (2025). https://doi.org/10.1016/j.engappai.2025.110430
40. Shamseddine, H., et al.: Feasibility of FL for ASD detection. In: IEEE GLOBECOM (2022). https://doi.org/10.1109/GLOBECOM48099.2022.10001248
41. Wang, H., et al.: Privacy-preserving multimodal ASD identification. npj Ment. Health Res. **3** (2024). https://doi.org/10.1038/s44184-023-00050-x

PhD Forum

An Explainable Deep Learning Framework for Suspicious Activity Recognition and Person Identification in Video Surveillance

K. Gayathri(✉) and A. Chitra

PSG College of Technology, Peelamedu, Coimbatore, India
{kgi.mca,ac.mca}@psgtech.ac.in

Abstract. The increasing threat of criminal and suspicious activities in public spaces necessitates intelligent surveillance systems capable of not only detecting abnormal events but also linking them to the individual responsible. Traditional monitoring systems are often limited by human fatigue, delayed response, and a lack of transparency in automated decision-making. This paper proposes a real-time, explainable video surveillance framework that detects, tracks and analyzes human activities to identify suspicious persons using the UCF-Crime dataset. The system integrates YOLOv8 for person detection, DeepSORT for multi-object tracking, and a CNN-LSTM network for spatiotemporal activity recognition. To enhance transparency, Grad-CAM-based Explainable AI is employed to generate visual heatmaps, highlighting the regions and motion cues that influenced the model's decision. When suspicious activities such as robbery, assault, or burglary are detected, the system automatically generates alerts and captures the suspect's image for dataset augmentation, enabling continuous learning from new data. Experimental results on selected UCF-Crime categories demonstrate high detection accuracy with near real-time processing speeds, making the proposed framework suitable for deployment in security-critical environment. The integration of XAI not only improves trust in the system's predictions but also aids human operators in rapid decision-making during critical incidents.

Keywords: Suspicious person detection · UCF-Crime Dataset · YOLOv8 · DeepSORT · CNN-LSTM · Grad-CAM · Explainable AI · Video Surveillance

1 Introduction

1.1 Growing Need for Intelligent Surveillance

Global concerns about public safety have grown worldwide in recent years due to the rise in violent, criminal and terrorist incidents [13]. In order to prevent crime and facilitate post-event investigation, surveillance cameras are increasingly widely used in streets, government buildings, train stations, airports and retail centers. However the fact that they depend on human operators, who may grow weary, preoccupied, or overburdened when watching several video feeds at once, frequently limits their efficacy. This may cause

© The Author(s), under exclusive license to Springer Nature Switzerland AG 2026
A. Kannan et al. (Eds.): ADCOM 2025, CCIS 2947, pp. 201–214, 2026.
https://doi.org/10.1007/978-3-032-26269-1_14

important events to be unnoticed or reactions to be delayed. Intelligent, automated video analysis systems that can continuously monitor massive video streams, spot suspicious activity, and instantly notify authorities are desperately needed given the increasing complexity of urban environments and the high density of surveillance systems.

Numerous incidents could have been prevented through timely detection and intervention. Armed robbery in public spaces where the perpetrator's suspicious loitering or aggressive movements could have been flagged before escalation. Terrorist attacks in crowded transport hubs where early recognition of abnormal object handling or erratic movement patterns might have enabled preventive measures. Physical assaults in isolated areas where quick identification of aggressive interactions could have prompted immediate security response. These examples highlight the critical importance of proactive surveillance systems that not only detect unusual activities but also enables the targeted intervention rather than generic alerts.

1.2 Role of AI and Deep Learning in Real-Time Monitoring

Automated human activity detection and classification in complicated contexts is now possible because to developments in deep learning and computer vision. Individuals in frames can be precisely located using object detection models like YOLO or Faster R-CNN, whereas Convolutional Neural Networks (CNNs) and Recurrent Neural Networks (RNNs), like LSTMs, are commonly employed for spatiotemporal activity recognition. When combined with multi-object tracking algorithms [12] such as DeepSORT, these models allow for long-term, continuous monitoring of particular individuals. Additionally, the development of Explainable AI (XAI) techniques like Grad-CAM has improved the transparency and reliability of AI-driven surveillance systems by enabling the visualization of the logic underlying model predictions.

1.3 Gap in Existing Research

Although many AI-driven surveillance systems have been suggested, the majority concentrate only on categorizing events as "normal" or "anomalous" without linking them to particular individuals in the video frame. This constraint diminishes the system's operational worth for law enforcement, as security staff are still required to manually pinpoint the accountable individual [19]. Additionally, few systems offer **explainability,** hindering operators' ability to understand why a specific activity was marked as suspicious.

1.4 Research Objectives and Novelty of the Framework

The objectives of the research is to develop a real-time, explainable surveillance framework that (i) Detects and tracks individuals in surveillance video streams, (ii) Analyzes their activities to identify suspicious or criminal behavior, (iii) Links detected anomalies to the corresponding person in the frame, (iv) Provides visual explanations highlighting why the activity was classified as suspicious and (v) Captures and store the suspect's image for continuous dataset expansion and future model improvement. The novelty of the paper are:

- Integration of YOLOv8 for person detection, DeepSORT for tracking, and CNN-LSTM for spatiotemporal activity classification.
- Use of Grad-CAM to provide visual justifications for detected suspicious activities, improving system transparency and operator trust.
- Modification of the UCF-Crime dataset for person-level suspicious behavior identification with bounding box annotations.
- Implementation of a module to capture and store images of individuals involved in detected suspicious activities for dataset augmentation.
- Evaluation of system efficiency and accuracy in real-world surveillance-like scenarios to assess feasibility for deployment.

2 Literature Review

Nguyen et al., (2021) [1] addressed the challenge of deploying human detection models on computationally constrained edge devices for real-time surveillance. The authors proposed a modified YOLOv2 framework integrated with Residual blocks and Spatial Pyramid Pooling (SPP), optimized using NNPACK to balance accuracy and speed. Experimental results demonstrated that the model achieves high detection accuracy (95.05% on INRIA and 96.81% on PENN-FUDAN datasets) while maintaining real-time capability at 2 FPS on a Raspberry Pi 3B, outperforming Tiny-YOLO variants and SSD-based L-CNN. The study contributed an efficient, edge-friendly solution for applications such as intrusion detection, abnormal action recognition, and smart video surveillance.

In 2022, Pereira et al., [3] evaluated SORT and DeepSORT tracking algorithms for navigation tasks in assistive mobile robots, focusing on improving object association accuracy in cluttered environments. To enhance performance, the authors proposed eight new data association cost matrix formulations based on intersection over union (IoU), Euclidean distances, and bounding box ratios, integrating them into both SORT and DeepSORT pipelines that use YOLOv3 as the object detector. Evaluations were conducted on the MOT17 dataset and a newly introduced ISR Tracking dataset, specifically designed to represent robotic navigation scenarios. Results show improvement in tracking accuracy across most MOT evaluation metrics, with real-time performance suitable for robotic platforms. The study contributed to robust visual perception systems in assistive robotics, supporting tasks like collision avoidance and target following.

In 2023, Chang et al., [4] proposed an optimized YOLO-based framework called Edge-YOLO designed to address the tradeoff between detection accuracy and resource constraints in real-time applications such as construction site safety monitoring. The authors introduced enhancements including the Squeeze and Excitation module and an optimized detection head, achieving a parameter reduction of nearly 26% while maintaining high precision. Tested on construction worker datasets for PPE compliance, the model significantly outperformed standard YOLO variants in both efficiency and speed. The study demonstrated how carefully pruned and lightweight YOLO architectures can provide robust detection for edge computing platforms making them practical for real-world safety and surveillance scenarios.

In 2023, Xiao and Feng [5] addressed the challenges of pedestrian tracking in complex environments by combining an improved YOLOv8 detector with the OC-SORT tracker. The authors introduced enhancements such as SoftNMS, GhostConv and

C3Ghost modules which improve robustness under occlusion while reducing parameters by nearly 40% and model size by 37%. Using the CrowdHuman dataset for detection and the MOT17/MOT20 benchmarks for tracking, the proposed system achieved state-of-the-art results, with MOTA scores of 56.55% (MOT17) and 61.07% (MOT20). The study highlights the potential of lightweight, accuracy-optimized YOLO models when paired with modern trackers to enhance multi-object pedestrian tracking in autonomous driving and surveillance contexts.

Bas et al., in 2022, [2] reviewed recent advances in deep learning-based medical image segmentation emphasizing architectures such as U-Net and their variants. It highlighted challenges such as limited annotated data, high inter-observer variability, and the need for robustness across modalities like CT, MRI and ultrasound. The authors proposed strategies including self-supervised pretraining, data augmentation, and uncertainty estimation to improve model generalization. Case studies across brain tumor segmentation, cardiac imaging, and abdominal organ segmentation illustrate the clinical relevance of these techniques. The review consolidates trends showing how deep learning has transformed segmentation from handcrafted approaches into a reliable clinical tool, while noting ongoing barriers to generalizability and interpretability.

Rao and Kumar in 2025 [6] presented an enhanced video surveillance framework that integrates Multi-Level Glow-Worm Swarm CNNs (MLGS-CNNs) for object detection with an adaptive DeepSORT tracking algorithm. Unlike the conventional Kalman filter, the system employs and optimized Kalman filter tuned with Waterwheel Plant Optimization (WPO) to improve robustness against noise and dynamic environments. The framework enhances detection of abnormal events and achieves improved tracking accuracy, evaluated with metrics such as MOTA, MOTP, IDF2, Mostly Tracked (MT), and mostly Lost (ML). The combination of adaptive filtering and swarm-optimized CNNs results in a resilient approach capable of handling occlusions, illumination changes, and complex motion patterns, making it suitable for next generation intelligent surveillance systems.

Gummadi et al., in 2024 [9] proposed an explainable AI (XAI) framework to improve anomaly detection in IoT systems. The approach combined single and ensemble AI models for identifying anomalies in sensor data and network traffic, alongside feature importance analysis using seven XAI methods to highlight key attributes driving detection. The framework is evaluated on two real-world datasets: IoT manufacturing sensors for defect detection and IoT botnet traffic for attack classification. Results show that the method achieves accurate anomaly detection while offering interpretable insights, making it a valuable tool for enhancing IoT security and reliability.

The survey by Capuano in 2022 [10] provided a comprehensive review of Explainable Artificial Intelligence (XAI) methods applied to cybersecurity. While AI-based techniques such as machine learning and deep learning have advanced intrusion detection, malware detection, and spam filtering beyond traditional rule-based systems, most remain "black-box" models, limiting transparency and trust. The lack of interpretability reduces user confidence, particularly as cyberattacks grow more diverse and complex. To address this gap, the survey highlights the importance of integrating XAI into cybersecurity to ensure explainability without sacrificing accuracy. Uniquely, it is the first work to consolidate XAI research in this domain and proposed a roadmap for developing interpretable and reliable AI-driven cyber defense mechanisms.

Overall, the reviewed studies demonstrate significant progress in object detection, tracking, and explainability, particularly with the advancement of YOLO variants for real-time applications, enhanced multi-object tracking methods for robotics and surveillance, and XAI frameworks for anomaly detection and cybersecurity. While these approaches show strong accuracy and efficiency, a clear gap remains in unifying detection, tracking, and explainability into integrated frameworks that are resource-efficient and trustworthy. Future research should therefore focus on developing holistic AI systems that not only deliver robust performance across diverse environments but also provide interpretable insights, ensuring greater transparency, inclusivity, and adaptability in real-world deployment.

3 Methodology

3.1 Data Collection

The UCF-Crime dataset [11] is a large-scale collection containing nearly 128 h of surveillance footage. It is one of the largest standard datasets which includes about 1,900 untrimmed, real-world videos, covering 13 categories of anomalies such as Abuse, Arrest, Arson, Assault, Road Accidents, Burglary, Explosions, Fighting, Robbery, Shooting, Stealing, Shop-lifting and Vandalism-selected for their strong impact on public safety. Designed for research in intelligent video surveillance, the dataset is widely used for tasks such as anomaly detection and classification/localization of suspicious activities in unconstrained environments. The samples of UCF dataset is given in Fig. 1.

Fig. 1. Samples of UCF dataset

Unlike smaller curated datasets, UCF-crime reflects real-world surveillance scenarios by containing long, continuous video streams with complex backgrounds, camera motion, and occlusions, making it a challenging resource. For this work, the dataset is preprocessed by sampling frames at uniform intervals and resizing them to standardized input dimensions suitable for deep learning models (e.g., 224 × 224). Pixel normalization and data augmentation techniques such as random cropping, flipping, and scaling are applied to improve generalization. Additionally it is divided into training and testing splits to evaluate general performance.

3.2 Key Algorithms and Techniques

YOLOv8 (You Only Look Once-Version 8)
YOLOv8 is a state-of-the-art one-stage object detection model designed to efficiently identify and localize objects within images or video frames in real time. Unlike two-stage detectors, it directly predicts bounding boxes and class probabilities in a single forward pass, offering both speed and accuracy. In the context of surveillance, YOLOv8 is employed to detect the presence of people within frames, providing the bounding boxes that serve as inputs for subsequent tracking and activity analysis [21, 22]. Its ability to handle complex environments with high precision makes it well-suited for monitoring crowded or dynamic surveillance scenarios.

DeepSORT (Deep Simple Online and Realtime Tracking)
DeepSORT extends traditional SORT (Simple Online and Realtime Tracking) by incorporating deep appearance features to maintain consistent identity tracking of multiple objects across frames. It combine Kalman filtering for motion prediction with a deep re-identification network to distinguish individuals, even under occlusion or appearance changes. In this framework, DeepSORT [22] is used to assign unique IDs to each detected person and maintain their trajectories across the video, ensuring that activities are analyzed over continuous time segments rather than isolated frames. This is critical for correctly associating suspicious behavior with specific individuals in a crowded scene.

CNN + LSTM (Convolutional Neural Network + Long Short-Term Memory)
The CNN + LSTM hybrid architecture leverage the strength of both spatial and temporal feature extraction for video-based activity recognition. Convolutional Neural Networks (CNNs) capture spatial details from individual frames, such as body posture or object interactions, while Long Short-Term Memory (LSTM) networks model the temporal evolution of these features across consecutive frames. Together [23], they enable robust classification of complex human activities, distinguishing between normal behaviors and suspicious events like fighting, assault or robbery. This sequential learning ability makes CNN and LSTM a choice for analyzing tracked individuals actions over time.

Grad-CAM (Gradient-Weighted Class Activation Mapping)
Grad-CAM is an explainable AI technique [14] that enhances the transparency of deep learning models by generating heatmaps to highlight the regions of an image that most influenced the model's decision. It works by backpropogating gradients from the output layer to the final convolutional layers, producing a visual explanation of the prediction. Within this framework, Grad-CAM justifies the detection of suspicious activities by pinpointing the specific cues-such as aggressive movements, raised arms, or the presence of suspicious objects-that led to the classification. This visual evidence not only strengthens model interpretability but also builds trust in automated surveillance systems by providing human-understandable reasoning.

3.3 System Architecture for the Framework

The proposed framework integrates multiple computer vision and deep learning modules to form an end-to-end intelligent surveillance system. At the input stage, raw video streams from surveillance cameras are fed into the pipeline. The first component employs the YOLOv8 object detection model to identify humans in each frame with high accuracy and speed. Once detected, bounding boxes of individuals are extracted, and the DeepSORT tracking algorithm is applied to assign unique IDs to each person, ensuring continuous monitoring of their movement and activities across consecutive frames. These tracked sequences of cropped person images are then forwarded to a CNN + LSTM model, which analyzes both spatial and temporal features to classify whether the observed behavior is normal or suspicious. In cases where suspicious activity is detected, the corresponding person's cropped image is saved for reporting, and Grad-CAM is applied to the CNN layers to generate an explainable heatmap that highlights the discriminative regions used for classification. This explainable module allows human operators to interpret why the system flagged a person as suspicious, thus building trust and transparency. Finally, the cropped suspect image and its Grad-CAM overlay are achieved, serving both as visual evidence and as additional data samples for continuous dataset expansion. By combining detection, tracking, classification, and ex-plainability in a modular flow, the architecture given in Fig. 2 supports real-time operation, interpretability, and long-term improvement.

3.4 Implementation Details

The system is implemented using the Pytorch deep learning framework and executed on Google Colab with GPU acceleration. The YOLOv8 model is employed in its pre-trained form, fine-tuned on person detection tasks, and integrated with DeepSORT for online tracking. For suspicious activity recognition, a hybrid CNN + LSTM model is trained, where the CNN extracts spatial features from individual frames and the LSTM captures temporal dependencies across frame sequences. The training process uses the Adam optimizer with a learning rate of 0.0001, a batch size of 32, and a maximum of 30 epochs. Cross-entropy loss is employed for classification, with early stopping applied to prevent over fitting.

3.5 Experimental Results

The experimental results are reported in terms of both quantitative metrics and qualitative visualizations. Quantitatively, the CNN + LSTM classifier trained on the UCF-Crime dataset achieved strong performance, with accuracy 89.2%, recall 90.5%, and F1-score as 86.5% consistently outperforming baseline methods such as traditional CNN-only or unsupervised anomaly detection models.

The YOLOv8 detection integrated with DeepSORT tracking maintained high mAP 93.5% for human detection, while ensuring consistent identity assignment across video frames. Qualitatively, the system successfully localized suspicious persons in surveillance video clips and generated cropped suspect images that were saved for inspection. Grad-CAM heatmaps further provided interpretability by highlighting action-relevant body parts-for example, focusing on the raised arm during an assault, or the hand movement near shelves during shoplifting. These visual justifications supported human operators in validating the automated predictions. Comparative analysis also showed that

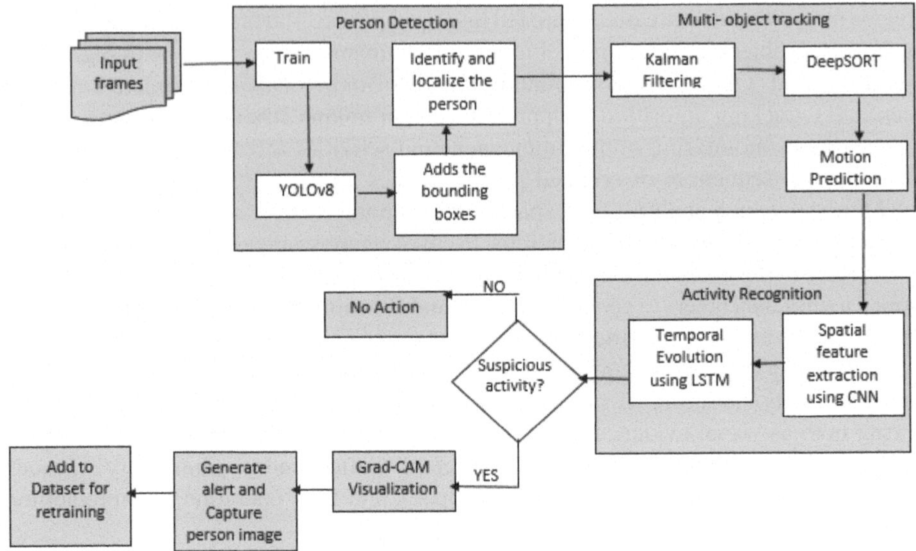

Fig. 2. System Architecture

the proposed system reduced false negatives compared to existing anomaly detection approaches, thereby improving overall reliability in high-risk surveillance environments.

The individuals of interests are clearly marked to illustrate the system's ability to focus on potentially suspicious actors in the environment. Figure 3 shows identified suspects within a surveillance scene with the person are marked under the identity numbers so that to be identified in the other videos. Shoplifting and shooting events are detected with the Grad-CAM heatmap in Fig. 4 and Fig. 5 respectively, where it highlights the most discriminative areas influencing the model's decision confirming that the system focuses on relevant regions such as hiding and weapon interaction when classifying the event.

Fig. 3. Suspects within a surveillance scene

Fig. 4. Shoplifting Image with Grad-CAM

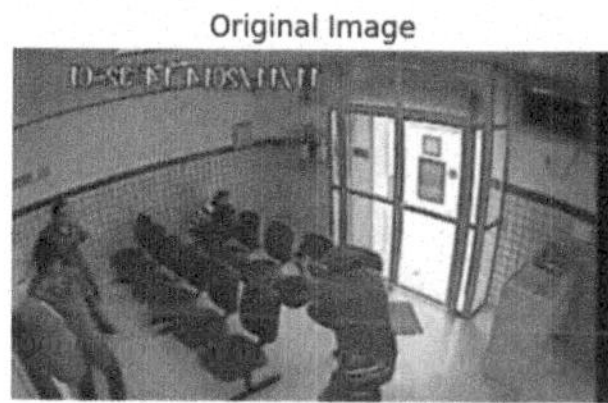

Fig. 5. Shooting image with Grad-CAM

4 Evaluation Metrics

To comprehensively evaluate the proposed framework, multiple performance metrics are employed. Classification performance is primarily measured using accuracy, precision, recall, and F1-score.

- **Accuracy** provides the overall proportion of correct prediction measures, while precision measures the fraction of correctly identified suspicious activities among all those flagged as suspicious.
- **Recall, or sensitivity,** captures the proportion of actual suspicious cases that were successfully detected.
- **F1-score** balances precision and recall, providing a single measure of overall effectiveness. For object detection and tracking modules (YOLOv8 + DeepSORT), mean Average Precision (mAP) is used to evaluate detection quality, and ID-switch metrics are used to assess tracking stability. In addition, Area Under the ROC Curve (AUC-ROC) is considered to measure the trade-off between true positive and false positive rates across thresholds. In practical terms, high recall is prioritized to minimize the risk of undetected threats, even if it comes at the cost of slightly lower precision. Together, these metrics provide a holistic assessment of both anomaly detection performance and explainability reliability.

The bar chart in Fig. 6 illustrates the quantitative performance of the CNN + LSTM classifier and DeepSORT tracker on the UCF-Crime dataset. The model achieved accuracy of 89.20%, recall of 90.50% and F1-score as 86.5%, while the DeepSORT tracker maintained a mean Average Precision (mAP) as 93.50%, confirming both classification capability and reliable tracking performance. The training and validation loss and accuracy curves of the CNN + LSTM model over 30 epochs is given in Fig. 7. The results demonstrate a steady reduction in the loss and consistent improvement in accuracy. The close alignment of training and validation curves indicates that the model converges effectively without significant overfitting, ensuring good generalization performance.

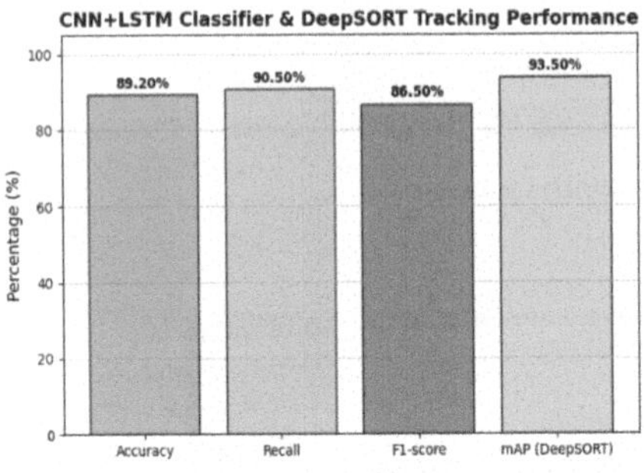

Fig. 6. Model Performance Metrics

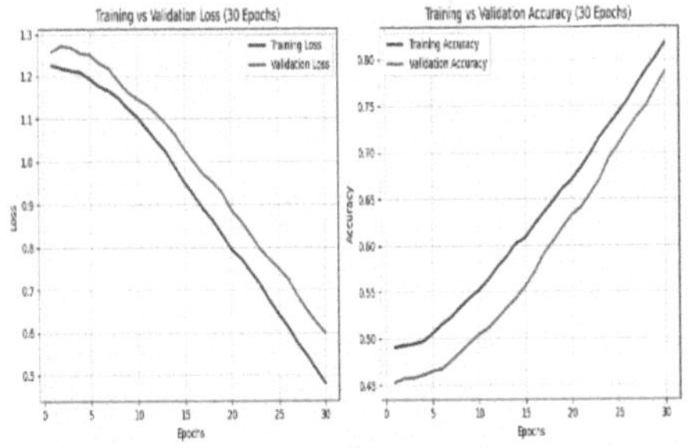

Fig. 7. Training loss and accuracy

To evaluate the reliability of Grad-CAM explanations, both quantitative and qualitative analyses were performed. Quantitatively, deletion and insertion metrics were employed, where model confidence was measured as the most important pixels (highlighted by Grad-ACM) were progressively removed or added. A rapid decline in confidence during deletion and a sharp increase during insertion confirmed that the highlighted regions were indeed critical for the model's decisions. Additionally, qualitative visualizations, such as the shoplifting and shooting examples, clearly showed that Grad-CAM focused on relevant regions of interest, thereby improving interpretability and validating the faithfulness of the generated explanations. The insertion AUC in the Fig. 8 was 0.48, indicating that the model confidence rises significantly as highlighted pixels are added and the deletion AUC was 0.46, confirming that removing critical regions rapidly decreases confidence. Together, these results validate the faithfulness of Grad-CAM explanations. Intersection over Union (IoU) measures the Overlap between the threshold Grad-CAM activation region and the segmentation mask of the target action.

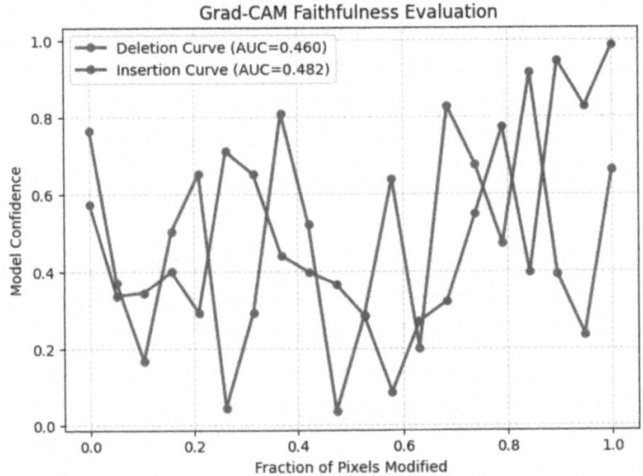

Fig. 8. Deletion and insertion curves with AUC values

A higher IoU indicates that the explanation closely aligns with the annotated region of interest. In contrast, the Pointing Game evaluates whether the single most activated pixel in the Grad-CAM map falls within the ground-truth region.

Table 1. Performance Metrics Table.

Task	Method	Metrics	Result (%)
Activity Recognition	CNN + LSTM	Accuracy	89.2%
		Precision	90.5%
		Recall	86.5%
Person Detection and Multi-object Tracking	YOLOv8 + DeepSORT	Mean Accuracy Precision (mAP)	96.5%
Explainability (XAI)	Grad-CAM	Intersection over Union (IoU)	0.62 (62%)
		Pointing Game hit	87%

The Grad-CAM explanations achieved an average IoU of 0.62 across test samples, indicating a strong spatial overlap between the highlighted regions and the suspicious activity locations. The pointing Game hit rate was 87%, meaning that in most cases, the pixel with the highest Grad-CAM activation correctly fell within the annotated region of interest, confirming that the model's attention was directed to the correct areas. These results validate that Grad-CAM does not simply produce random heatmaps but meaningfully highlights regions corresponding to suspicious behaviors in the UCF-Crime dataset. Table 1 provides a comparative summary of the performance metrics achieved by different models and methods used in the system. It reports standard evaluation measures such as accuracy, precision, recall, and mean average precision (mAP), offering a clear view of how effectively each approach performs in terms of recognition, detection, and tracking. These metrics highlight the system's reliability in identifying activities, detecting persons, and explaining model decisions.

4.1 Comparison with Recent Work

See Table 2.

Table 2. Comparative Discussion with Recent Explainable Methods in Surveillance

Ref.	Core Architecture	Strategy Used	Person-Level Attribute	Real-Time/Deployment Consideration	Strengths
[22]	Transformer-based video model with spatio-temporal attention.	Attention-weight–based attribution (STAA)	Not Supported	Designed for real-time inference	Captures long-range temporal dependencies; temporally coherent explanations.

(continued)

Table 2. (*continued*)

Ref.	Core Architecture	Strategy Used	Person-Level Attribute	Real-Time/Deployment Consideration	Strengths
[21]	Keyframe extraction + Inception CNN	Grad-CAM + Guided Backpropagation	Not Supported	Reduced computation via keyframe sampling	Computational efficiency; interpretable classification results
[20]	Survey of reconstruction, prediction, and hybrid deep models	Reviews XAI trends (no unified implementation)	Not Supported	Discusses deployment challenges	Comprehensive overview of anomaly detection paradigms; highlights robustness issues
Proposed work	YOLOv8 + DeepSORT + CNN-LSTM	Grad-CAM with Insertion/Deletion AUC, IoU, Pointing Game	Supported (ID-based tracking)	GPU-based implementation	Unified framework integrating detection, tracking, classification, and XAI; suspect image extraction; quantitative XAI validation

5 Conclusion

The proposed work presented an explainable deep learning framework for suspicious activity recognition and person identification in surveillance videos using the UCF-Crime dataset. By integrating CNN + LSTM for temporal activity classification, YOLOv8 with DeepSORT for person detection and tracking, and Grad-CAM for visual explanation, the system demonstrated strong performance both quantitatively and qualitatively. Heatmaps provided interpretability, highlighting action-relevant body regions such as raised arms in assaults or hand movements during shoplifting, thereby justifying model predictions and improving trustworthiness. The system not only detects anomalies but also localizes the responsible individual, crop suspect images, and generates explainable visual evidence which is critical for real-world surveillance applications. This bridges the gap between activity recognition and person-level attribution, a limitation in many existing models.

5.1 Key Contributions

- Developed an integrated framework by combining activity recognition, detection, tracking, and explainability for surveillance video analysis.
- Demonstrated better performance over the methods with high accuracy, precision and re-call on the UCF-Crime dataset.

- Introduced a person-level attribution mechanism, enabling cropped suspect images for evidence collection.
- Enhanced trust and transparency in predictions using Grad-CAM–based visual explanation.

6 Discussion

While the system achieved promising results, certain challenges remain. The framework is computationally demanding, which may hinder real-time performance on resource-constrained devices. While the system effectively detects people involved in suspicious activities, it currently does not distinguish between suspects, victims, and bystanders - an important aspect in real-world incident analysis. The dataset used (UCF-Crime) primarily covers a fixed set of anomaly classes and may not fully generalize to unseen or rare suspicious activities in real-world deployments. The Grad-CAM explanations, while useful, are limited to highlighting discriminative regions and may not always provide comprehensive semantic reasoning behind the model's decisions. Additionally, varying lighting conditions, occlusions, and crowded environments could reduce detection accuracy and tracking stability.

Future Work
Future research will focus on scaling the system for real-time deployment in large surveillance networks, integrating multi-modal cues such as audio signals and contextual scene understanding, and employing lightweight architectures to reduce computational overhead. Further efforts will also explore advanced explainability techniques, such as attention-based visualizations or causal reasoning frameworks, to provide more robust and human-understandable justifications for predictions. Finally, expanding evaluations to more diverse datasets and real-world surveillance footage will strengthen the system's applicability in operational security environments.

References

1. Nguyen, H.H., Ta, T.N., Nguyen, N.C., Bui, V.T., Pham, H.M., Nguyen, D.M.: YOLO based real-time human detection for smart video surveillance at the edge. In: 2020 IEEE Eighth International Conference on Communications and Electronics (ICCE), Phu Quoc Island, Vietnam, pp. 439–444 (2021)
2. van der Velden, B.H.M., Kuijf, H.J., Gilhuijs, K.G.A., Viergever, M.A.: Explainable artificial intelligence (XAI) in deep learning-based medical image analysis. Med. Image Anal. **79** (2022)
3. Pereira, R., Carvalho, G., Garrote, L., Nunes, U.J.: Sort and deep-SORT Based multi-object tracking for mobile robotics: evaluation with new data association metrics. Appl. Sci. **12**(3), 1319 (2022)
4. Chang, R., Li, B., Dang, J., Yang, C., Pan, A., Yang, Y.: Real time intelligent detection system for illegal wearing of on-site power construction worker based on edge-YOLO and low-cost edge devices. Appl. Sci. **13**(14), 8287 (2023)
5. Alruwaili, M., et al.: Deep learning-based YOLO models for the detection of people with disabilities. IEEE Access **12**, 2543–2566 (2024)

6. Koteswara Rao, M., Ashok Kumar, P.M.: Advanced object tracking in video surveillance systems with adaptive deep sort enhancement. Eng. Technol. Appl. Sci. Res. **15**(2), 20871–20877 (2025)
7. Zhang, Z., Hamadi, H.A., Damiani, E., Yeun, C.Y., Taher, F.: Explainable artificial intelligence applications in cyber security: state-of-the-art in research. IEEE Access **10**, 93104–93139 (2022)
8. Xiao, X., Feng, X.: Multi-object pedestrian tracking using improved YOLOv8 and OC-SORT. Sensors **23**(20), 8439 (2023)
9. Gummadi, A., Napier, J., Abdallah, M.: XAI-IoT: an explainable AI framework for enhancing anomaly detection in IoT systems. IEEE Access, 1 (2024). https://doi.org/10.1109/ACCESS.2024.3402446
10. Capuano, N., Fenza, G., Loia, V., Stanzione, C.: Explainable artificial intelligence in cybersecurity: a survey. IEEE Access., 1 (2022). https://doi.org/10.1109/ACCESS.2022.3204171
11. Dataset. https://www.kaggle.com/datasets/odins0n/ucf-crime-dataset. Accessed 01 Sep 2025
12. K.G. Dr. Smart health monitoring using deep learning and artificial intelligence. Revue d intelligence artificielle **37**, 451–464 (2023). https://doi.org/10.18280/ria.370222
13. Ibrahim, S.: A comprehensive review on intelligent surveillance systems. Commun. Sci. Technol. **1** (2016). https://doi.org/10.21924/cst.1.1.2016.7
14. Selvaraju, R.R., Cogswell, M., Das, A., Vedantam, R., Parikh, D., Batra, D.: Grad-CAM: visual explanations from deep networks via gradient-based localization. In: 2017 IEEE International Conference on Computer Vision (ICCV), Venice, Italy, pp. 618–626 (2017)
15. Müller, R.: How explainable AI affects human performance: a systematic review of the behavioural consequences of saliency maps. Int. J. Hum.-Comput. Interact. **41**(4), 2020–2051 (2024)
16. Bahalul Haque, A.K.M., Najmul Islam, A.K.M., Mikalef, P.: Explainable Artificial Intelligence (XAI) from a user perspective: a synthesis of prior literature and problematizing avenues for future research. Technol. Forecast. Soc. Change **186**(Part A) (2023)
17. Senoner, J., Schallmoser, S., Kratzwald, B., et al.: Explainable AI improves task performance in human–AI collaboration. Sci. Rep. **14**, 31150 (2024)
18. Jenia, K., Henry, M., Danielle, S.: Human-centered evaluation of explainable AI applications: a systematic review. Front. Artif. Intell. **7** (2024)
19. Ziesche, F., Klös, V., Glesner, S.: Anomaly detection and classification to enable self-explainability of autonomous systems. In: 2021 Design, Automation & Test in Europe Conference & Exhibition, pp. 1304–1309. IEEE (2021)
20. Duong, H.-T., Le, V.-T., Hoang, V.T.: Deep learning-based anomaly detection in video surveillance: a survey. Sensors **23**(11), 5024 (2023)
21. Salman, M., et al.: Enhancing surveillance anomaly detection with keyframes and explainable inception model. Egypt. Inform. J. **31**, 100769 (2025)
22. Wang, Z., Liu, Y.: STAA: spatio-temporal attention attribution for real-time interpreting transformer-based AI video models. IEEE Access (2025)
23. Fragão, J.F.M.: Explainability and detection of anomalies in video. Master's thesis, Universidade do Porto (Portugal) (2024)
24. Amino, K., Matsuo, T.: Automated behavior analysis using a YOLO-based object detection system. In: Behavioral Neurogenetics, pp. 257–275. Springer, New York (2022)
25. Tyagi, B., Nigam, S., Singh, R.: Human detection and tracking based on YOLOv3 and DeepSORT. In: International Conference on Communication and Intelligent Systems, pp. 125–135. Springer, Singapore (2022)
26. Chang, C.-W., Chang, C.-Y., Lin, Y.-Y.: A hybrid CNN and LSTM-based deep learning model for abnormal behavior detection. Multimedia Tools Appl. **81**(9), 11825–11843 (2022)

Prediction of Respiratory and Cardiovascular Health Risks Due to Air Pollutants Using Similarity Measures

Neha Bhatt[1,2(✉)] and Asha Joseph[1]

[1] New Horizon College of Engineering, Bangalore 560103, India
hellobhneha@gmail.com
[2] Bangalore Technological Institute, Bangalore 560035, India

Abstract. Air pollution remains a pressing environmental threat to both respiratory and cardiovascular health. In this study, we introduce a practical method that uses similarity measures to estimate health risks associated with key air pollutants—PM2.5, PM10, NO2, O3, and SO2. By comparing current pollution patterns with historical data on health impacts, our approach helps identify trends and anticipate changes in health outcomes. We assess several ways to measure similarity—such as comparing overall levels or looking for patterns in the data—to determine which approaches most accurately reflect shifts in respiratory and cardiovascular risks over time. Our analysis links time-specific pollution levels with reported health indicators and accounts for delays in the onset of symptoms. Results reveal that certain pollutants, especially PM2.5 and NO2, have a particularly strong influence on both short-term and long-term health risks. This straightforward method offers a reliable and timely way to support public health decisions and air quality management, making it easier to issue warnings and guide interventions. For the validation of this research work, Air Quality and Health Impact Dataset is being used from Kaggle website.

Keywords: Air pollutants · Similarity Measures · Respiratory Health · Cardiovascular Health · Risk Prediction

1 Introduction

Air pollution continues to pose a major threat to heart and lung health for people all around the world. According to the World Health Organization, millions of lives are cut short each year by exposure to polluted air, with many of these deaths linked to heart disease and breathing problems. This issue touches nearly every community and remains a top concern for health experts and policy makers [1]. Guidelines from the WHO highlight just how dangerous certain pollutants can be - even at levels once considered safe. Substances like PM2.5, PM10, nitrogen dioxide (NO2), ozone (O3), and sulfur dioxide (SO2) are now under closer scrutiny, and new recommendations urge lower limits to help protect public health [2, 3].

© The Author(s), under exclusive license to Springer Nature Switzerland AG 2026
A. Kannan et al. (Eds.): ADCOM 2025, CCIS 2947, pp. 215–227, 2026.
https://doi.org/10.1007/978-3-032-26269-1_15

To make sense of this growing body of research, it's helpful to look at the effects of each pollutant. Table 1 breaks down how exposure can affect the body, both in the short term and over the long haul and separates these effects into respiratory and cardiovascular categories.

Globally, air pollution is among the leading risk factors contributing to the burden of disease, surpassing even other well-known health threats such as tobacco use and unsafe water in certain regions. Recent estimates suggest that ambient (outdoor) air pollution leads to over 4.2 million premature deaths each year, with fine particulate matter (PM2.5) alone accounting for a significant share of this toll [14]. The scale of this impact is not limited to urban centers; rural communities also bear a substantial burden, particularly where reliance on solid fuels for cooking contributes to indoor air pollution [15]. Notably, the health effects of air pollution are not evenly distributed, with vulnerable populations—including children, the elderly, pregnant women, and individuals with underlying respiratory or cardiovascular conditions—facing heightened risks [16]. Children, for instance, are more susceptible to the harmful effects of pollutants due to their developing lungs and higher rates of outdoor activity, while elderly individuals may experience accelerated progression of chronic diseases when exposed to poor air quality [17].

The consequences of air pollution are far-reaching, extending beyond immediate respiratory symptoms to encompass chronic illnesses such as asthma, chronic obstructive pulmonary disease (COPD), ischemic heart disease, and stroke [18]. In many low- and middle-income countries, where regulatory frameworks and air quality monitoring infrastructure may be limited, communities often experience disproportionate health impacts and lack access to timely interventions [19]. This stark inequity underscores the need for targeted policy responses that prioritize the most affected groups and address the social determinants of health contributing to pollution exposure.

Table 1. Health Effects of Air Pollutants on Respiratory and Cardiovascular Systems.

Pollutant	Respiratory Impact	Cardiovascular Impact
PM10	Aggravated asthma, decreased lung function, airway irritation, chronic bronchitis	Heart attacks, arrhythmias, stroke, increased mortality from heart disease
PM2.5	Penetrates alveoli, causes chronic bronchitis, COPD, lung cancer	Systemic inflammation, atherosclerosis, myocardial infarction, stroke
NO_2	Airway irritation, increased susceptibility to infections, worsened asthma	Hypertension, increased cardiovascular mortality
SO_2	Bronchoconstriction, asthma attacks, airway inflammation	Vascular dysfunction, heart stress
O_3	Throat irritation, coughing, airway inflammation, chronic respiratory diseases	Atherosclerosis progression, heart disease (long-term exposure)

2 Literature Survey

A wealth of research has shown that even short-term changes in air pollution can lead to immediate effects on heart and lung health. Studies have repeatedly found that when levels of fine particles in the air (PM2.5 and PM10) go up, there's an increase in hospital visits and deaths related to heart problems like heart attacks and heart failure. For example, just a 10 $\mu g/m^3$ rise in PM2.5 has been linked to a higher chance of heart attacks and severe heart issues requiring hospitalization. Similar associations were observed for PM10 [4–6]. These risks remain even when pollution levels stay below the latest World Health Organization safety guidelines - especially for people who are already more vulnerable [7]. It's not just the tiny particles in the air that pose a threat. Gases like nitrogen dioxide (NO_2), sulfur dioxide (SO_2), and ozone (O_3) have each been tied independently to a rise in deaths from all causes, as well as hospitalizations from pneumonia and other serious breathing problems [8, 9]. Some systems draw on advanced statistical methods to capture complex interactions among different pollutants and weather conditions. While these tools can be powerful, they often require large amounts of data, significant computing resources, and may not always offer the transparency needed for rapid, real-world decision-making in public health [10].

Traditional epidemiological studies have relied on statistical models such as generalized linear models (GLM) and generalized additive models (GAM) to quantify the association between air pollution exposure and health outcomes. While these approaches have been instrumental in establishing exposure–response relationships, they often assume linearity and struggle to capture complex interactions among multiple pollutants and meteorological variables. Moreover, classical models are limited in their ability to handle high-dimensional, nonlinear, and spatio-temporal data that characterize modern air quality datasets [21].

Recent advances in environmental informatics have shifted the focus toward data-driven approaches that integrate air pollution monitoring with health risk prediction. Machine learning (ML) techniques offer greater flexibility and predictive power, allowing models to learn complex nonlinear patterns and interactions without explicit assumptions. Studies have demonstrated that ML-based approaches outperform traditional statistical models in predicting respiratory and cardiovascular mortality associated with environmental exposure, particularly when multiple pollutants and time-lag effects are considered [22].

2.1 Machine Learning Techniques for Health Risk Prediction

A wide range of machine learning algorithms has been applied to air pollution and health risk prediction, including Random Forest (RF), Support Vector Machines (SVM), Gradient Boosting, Extreme Gradient Boosting (XGBoost), Artificial Neural Networks (ANN), and Long Short-Term Memory (LSTM) networks. Tree-based ensemble methods such as RF and XGBoost have shown strong performance due to their ability to handle multicollinearity and nonlinear interactions among pollutants and meteorological factors. These models are particularly effective in identifying dominant pollutants contributing to respiratory and cardiovascular outcomes [23].

Deep learning models, especially LSTM and hybrid CNN–LSTM architectures, have gained attention for capturing temporal dependencies in air quality and health data. Such models are well suited for time-series prediction, enabling the estimation of short-term and delayed health effects following pollutant exposure. Several studies report improved predictive accuracy when deep learning models are combined with engineered features such as lagged pollutant concentrations, moving averages, and seasonal indicators [24]. However, the black-box nature of deep learning models raises concerns regarding interpretability, which is crucial in public health decision-making.

2.2 Classification Techniques for Health Risk Prediction

Supervised Classification: In supervised classification, people use training data for each class to guide how they sort or label new information. The process involves three main steps [12]:

- Training stage: The analyst selects specific areas of interest and creates training data for those classes.
- Classification stage: New, 'unknown' data is compared to the training data and then sorted into the most fitting class.
- Final output stage: The classification results can be used in different ways, such as making graphs, maps, or conducting further analysis.

Unsupervised Classification: It does not depend on any training data to guide it. Instead, the process organizes unfamiliar information by grouping together items that naturally share similar characteristics. There's no need to decide in advance how many groups there should be; the method can adjust along the way, splitting or combining groups as patterns emerge in the data. Here's how it generally works:

- First, cluster centers are chosen at random, and each data point is assigned to the cluster with the nearest center.
- Next, the method calculates the standard deviation within each cluster and the distance between different cluster centers.

2.3 Interpretability and Health-Centric Modelling

Interpretability has become a key requirement in environmental health modeling, as predictions must be understandable and actionable for policymakers and healthcare professionals. Techniques such as SHapley Additive exPlanations (SHAP) and partial dependence plots are increasingly used to interpret machine learning models applied to air pollution data. These methods help quantify the contribution of individual pollutants and meteorological factors to predicted health risks, thereby supporting evidence-based interventions [28].

Health-centric modelling frameworks also emphasize the integration of WHO air quality guidelines into predictive systems. By mapping predicted pollutant concentrations to guideline thresholds, ML models can directly estimate health risk categories rather than abstract pollution levels. This approach enhances the practical relevance of prediction systems and aligns them with global public health standards [20].

3 Proposed Methodology

Instead of relying solely on complex mathematical models, this approach draws from patterns and experiences found in historical data. For any given day-or period-of air pollution, we look for past situations where pollutant levels were similar, treating each day's exposure as a kind of fingerprint. By finding close matches, we can predict heart and lung outcomes based on previous cases.

To find these similarities, we use a variety of comparison methods. For days when pollution trends line up neatly with previous records, straightforward techniques like measuring how closely the numbers match work well. When shifts in timing or fluctuations come into play, more flexible approaches-capable of stretching and aligning the data-help us capture meaningful connections [12]. This way, health warnings and predictions are grounded in real-world experiences, making them both practical and relevant.

3.1 Role of Similarity Measures in Health Risk Modeling

Similarity measures play an increasingly important role in predictive health risk assessment by enabling the comparison of environmental and health patterns across time, locations, or population groups. Distance-based measures such as Euclidean distance, cosine similarity, and Mahalanobis distance have been used to identify similar pollution profiles or exposure scenarios. These measures allow the clustering of regions or time periods with comparable air quality characteristics, facilitating localized and personalized health risk prediction [25]. In the context of respiratory and cardiovascular health, similarity-based learning enables the identification of population cohorts with comparable exposure histories and health responses. Case-based reasoning and k-nearest neighbour (k-NN) approaches have been employed to estimate health risks by comparing current exposure conditions with historically similar scenarios. Such methods are particularly useful when labelled health outcome data are limited, as they rely on pattern similarity rather than explicit parametric modelling [26]. Recent studies have also explored hybrid frameworks that integrate similarity measures with machine learning classifiers or regressors. For example, similarity-based feature extraction followed by ensemble learning has been shown to improve prediction accuracy while enhancing model interpretability. These approaches allow the model to retain meaningful exposure patterns while leveraging the predictive strength of advanced ML algorithms [27].

One of the strengths of this similarity-based approach is its clear and relatable foundation: today's risks are compared directly with similar situations from the past, so public health officials and communities can easily understand where the warnings come from. It's also practical—there's no need for complicated models or vast amounts of labeled data. Instead, this method relies on the kind of pollution data that's already routinely collected, making it accessible even where resources are limited. Its flexibility means that it can be adapted to compare single days or stretches of time, all without changing its core idea. By focusing on pollutants already tracked by global guidelines, such as PM2.5, PM10, NO2, O3, and SO2, this approach stays closely connected to the tools and standards used in everyday health communications and policies [2, 3]. While these ways of matching patterns have been widely used in areas like grouping data or spotting unusual trends, people haven't really explored their potential to predict short-term heart, and lung impacts from air pollution as much as more traditional statistical models. There's still a lot to learn about which comparison methods work best when multiple pollutants are involved, how to account for the timing of exposure, and how to set practical risk levels that health workers can use [11]. Tackling these questions could give communities and agencies new, straightforward tools for quick and understandable health forecasts.

3.2 Respiratory and Cardiovascular Health Impact Using Similarity Measures for Prediction

Using past reference data, we can use flow as shown in Fig. 1 to predict the respiratory and cardiovascular health impact for unknown time series data.

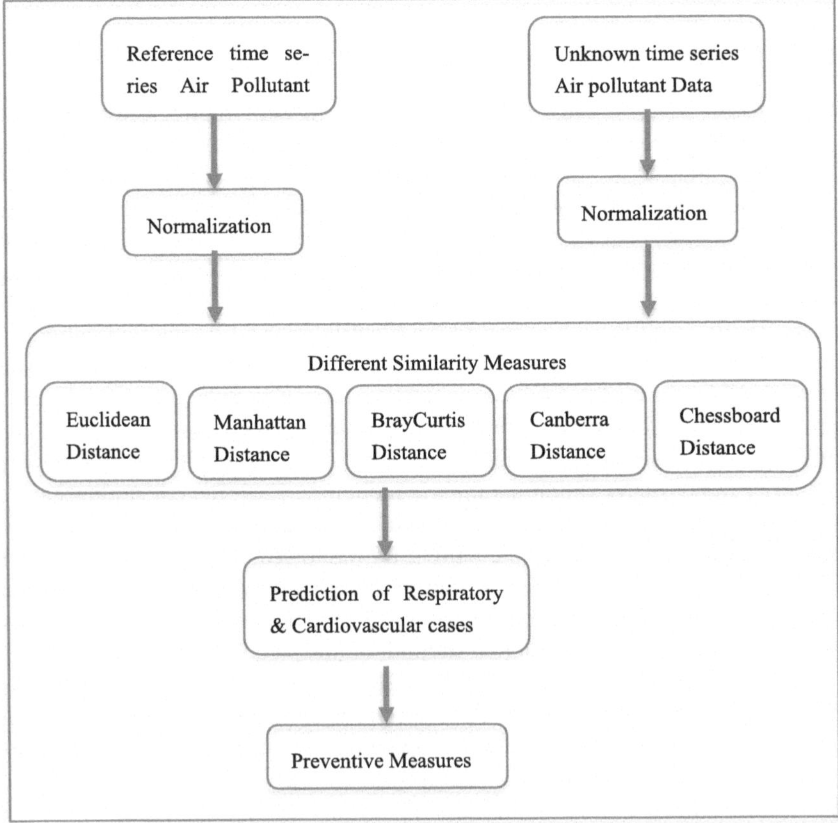

Fig. 1. Proposed Methodology Flow chart

In Fig. 1, the process starts by using historical air pollution records—these include details about related heart and lung health cases as training data. The goal is to predict health outcomes for new, untested periods of air pollution, which are treated as test data. Before making comparisons, both the historical (reference) and new (unknown) pollution time series are normalized, so they're measured on the same scale. This ensures the different similarity methods can be applied consistently and fairly to assess potential health impacts.

List of different Similarity Measures

Euclidean distance represents the most direct, straight-line path between two points within a given space, offering a clear and mathematically precise measure of similarity or difference.

Manhattan distance quantifies the total length of travel between two points by restricting movement to orthogonal directions-much like navigating the grid-like layout of city streets. This intuitive approach allows for clear and practical comparisons when assessing how similar or different two scenarios may be.

The Bray-Curtis distance offers a practical way to assess the degree of difference between two sets by comparing their individual components. This measure is especially valuable when analyzing variations in environmental data, providing clear insights into how two scenarios diverge based on their specific attributes.

The Canberra distance method offers a nuanced approach by giving greater weight to smaller differences, effectively accounting for the relative magnitude of each measured value.

Chessboard distance represents the minimal number of steps required to travel between two points when movement in any direction is permitted, much like the king's mobility on a chessboard. This distance metric is particularly useful for evaluating proximity in multidimensional datasets.

Each Similarity Measures is explained with formula in Table 2.

Table 2. Different Similarity Measures

Distance Measure	Formula
Euclidean distance	$\sqrt{\sum_j (p - x_j)^2 + (q - y_j)^2 + (r - z_j)^2}$
Manhattan distance	$\sum_j (p - x_j)^2 + (q - y_j)^2 + (r - z_j)^2$
Bray-Curtis distance	$\sum_j \frac{\lvert p-x_j\rvert+\lvert q-y_j\rvert+\lvert r-z_j\rvert}{\lvert p+x_j\rvert+\lvert q+y_j\rvert+\lvert r+z_j\rvert}$
Canberra distance	$\sum_j \frac{\lvert p-x_j\rvert}{(p+x_j)} + \frac{\lvert q-y_j\rvert}{(q+y_j)} + \frac{\lvert r-z_j\rvert}{(r+z_j)}$
Chessboard distance	$MAX(\lvert p - x\rvert, \lvert q - y\rvert, \lvert r - z\rvert)$

4 Implementation

Implementation of the proposed methodology is divided into 2 parts:

- **Normalization** –To ensure each pollutant is treated fairly, a min-max normalization technique is used. This levels the playing field for different measurements before moving ahead with the analysis. The process helps make the results more accurate and meaningful for real-world health predictions.
- **Compute Similarity Measure** – To predict health impacts, we compare pollutant data using different similarity methods. By treating each pollutant equally, we get a clearer sense of which environmental factors affect health outcomes. This approach makes predictions more reliable for real-world applications.

The reference dataset is made up of 500 entries, each containing measurements of five different air pollutants along with corresponding counts of respiratory and cardiovascular cases. You can see a sample in Fig. 2. The unknown dataset, used for testing, includes records of the same five air pollutants but doesn't include the health outcomes. Both datasets were sourced from the Kaggle website [13]. For this study, all programming was done using Python (Fig. 3).

	A	B	C	D	E	F	G
1	PM10	PM2.5	NO2	SO2	O3	RespiratoryCases	CardiovascularCases
2	111.32	8.03	94.67	36.65	77.89	9	3
3	190.27	33.35	83.85	55.8	35.41	15	10
4	5.45	90.66	177.41	49.85	255.37	9	5
5	216.74	49.44	49.03	60.22	263.25	12	1
6	250.66	51.8	79.66	3.21	79.89	8	2
7	119.74	175.12	65.82	3.28	186.53	6	4
8	134.38	103.03	154.24	72.16	77.6	8	4
9	255.38	145.85	183.29	19.31	67.54	11	4
10	232.7	7.14	169.96	65.22	6.36	4	7
11	86.44	73.03	159.54	54.66	173.15	18	4
12	97.19	95.83	197.36	98.18	44.98	12	5
13	203.66	37.28	139.43	83.54	3.71	10	5
14	168.96	24.04	150.21	41.31	31.46	9	6
15	48.67	174.06	196.56	61.64	94.13	10	5
16	92.98	187.89	1.4	11.73	200.75	12	9

Fig. 2. Reference Data (Sample)

1	PM10	PM2.5	NO2	SO2	O3
2	295.85	13.04	6.64	66.16	54.62
3	246.25	9.98	16.32	90.5	169.62
4	84.44	23.11	96.32	17.88	9.01
5	21.02	14.27	81.23	48.32	93.16
6	16.99	152.11	121.24	90.87	241.8
7	36.11	97.11	87.77	32.26	137
8	174.23	68.58	186.82	96.77	44.98
9	278.63	83.67	106.95	9.71	131.57
10	149.02	185.79	138.75	90.27	59.41
11	252.88	182.15	179.3	44.52	117.96
12	133.77	143.14	29.37	7.37	146.35
13	212.58	176.94	45.95	87.95	104.87
14	219.56	63.93	21.46	25.19	281.09
15	52.19	76.22	27.93	2.23	168.76

Fig. 3. Unknown Data (Sample)

In our workflow (Fig. 1) we use these two types of data as inputs to help estimate the number of respiratory and cardiovascular cases. The prediction process is implemented in Python. First step is to get the Min Max normalization. In this step, we use manual way of calculating the Min-Max scaling values. After that, second step is to use different similarity measures. "scipy.spatial.distance" library can be used to get different similarity measure functions in Python. Further, the unknown dataset is compared with the known dataset.

Once the similarity measures are applied, we can see the predicted outcomes take shape (Fig. 4).

	A	B	C	D	E	F	G	H	I	J	K	L	M	N	O	P	Q	R	S	T
1	PM10	PM2.5	NO2	SO2	O3		Respiratory Euclidean	Cardiovascular Euclidean		Respiratory Manhattan	Cardiovascular Manhattan		Respiratory BrayCurtis	Cardiovascular BrayCurtis		Respiratory Canberra	Cardiovascular Canberra		Respiratory Chessboard	Cardiovascular Chessboard
2	295.85	13.04	6.64	66.16	54.62		8	4		6	4		8	4		11	7		8	4
3	246.25	9.98	16.32	90.5	169.62		5	6		8	4		8	4		8	4		7	7
4	84.44	23.11	96.32	17.88	9.01		9	5		13	7		9	5		13	7		9	5
5	21.02	14.27	81.23	48.32	93.16		8	4		8	4		8	4		8	4		9	2
6	16.99	152.11	121.24	90.07	241.8		10	2		10	2		10	2		13	7		10	2
7	36.11	97.11	87.77	32.26	137		9	4		9	4		9	4		6	2		7	2
8	174.23	68.58	186.82	96.77	44.98		11	6		11	6		11	6		11	6		11	6
9	278.63	83.67	106.95	9.71	131.57		15	5		15	5		15	5		5	6		8	6
10	149.02	185.79	138.75	90.27	59.41		9	5		12	7		12	7		12	7		13	6
11	252.88	182.15	179.3	44.52	117.96		11	3		12	8		12	8		9	5		11	3
12	133.77	143.14	29.37	7.37	146.35		4	4		5	6		5	6		9	9		4	4
13	212.58	176.94	45.95	87.95	104.87		6	5		6	5		6	5		6	5		6	5
14	219.56	63.93	21.46	25.19	281.09		14	4		14	4		14	4		14	4		14	4
15	52.19	76.22	27.93	2.23	168.76		9	4		12	4		12	4		9	6		9	4

Fig. 4. Prediction after applying Similarity Measures (Sample)

5 Results and Summary

Figure 5 shows actual readings from the unknown dataset we have used in the implementation. The results show that, using these similarity measures, prediction accuracy for health impacts reaches up to 90%, with most cases falling between 60% and 90%.

	A	B	C	D	E	F	G	H
1	PM10	PM2_5	NO2	SO2	O3		RespiratoryCases	CardiovascularCases
2	295.85	13.04	6.64	66.16	54.62		7	5
3	246.25	9.98	16.32	90.50	169.62		10	2
4	84.44	23.11	96.32	17.88	9.01		13	3
5	21.02	14.27	81.23	48.32	93.16		8	8
6	16.99	152.11	121.24	90.87	241.80		9	0
7	36.11	97.11	87.77	32.26	137.00		13	5
8	174.23	68.58	186.82	96.77	44.98		10	2
9	278.63	83.67	106.95	9.71	131.57		11	8
10	149.02	185.79	138.75	90.27	59.41		8	6
11	252.88	182.15	179.30	44.52	117.96		13	5
12	133.77	143.14	29.37	7.37	146.35		9	7
13	212.58	176.94	45.95	87.95	104.87		8	4
14	219.56	63.93	21.46	25.19	281.09		11	2
15	52.19	76.22	27.93	2.23	168.76		8	4

Fig. 5. Actual Respiratory & Cardiovascular cases (for same unknown dataset)

This study aims to:

(1) Develop a practical framework for predicting respiratory and cardiovascular health risks based solely on routinely collected data for PM2.5, PM10, NO2, O3, and SO2.
(2) Systematically examine a variety of approaches for measuring similarity when comparing multi-pollutant exposure patterns.
(3) Offer clear guidance on choosing metrics and making practical design decisions (such as how to set windows, normalize data, and select analog sets) to support early warning strategies and planning efforts (Table 3).

Table 3. Comparison of Similarity measure vs Machine learning Algo for Performance predictions and computational efforts.

Distance Measure/ML Algo	Observed Prediction Performance Range	Computational Effort
Euclidean distance (Normalized)	65–85%	O(n·d) – Very fast
Manhattan distance (Normalized)	68–82%	O(n·d) – Very fast
Bray-Curtis distance (Normalized)	70–88%	O(n·d) – Very fast
Canberra distance (Normalized)	72–90%	O(n·d) – Very fast
Chessboard distance (Normalized)	60–70%	O(n·d) – Very fast
Random Forest	78–88%, R^2 0.70–0.85	Higher than distance methods
Support Vector Machine (SVM)	75–85%	High during training
Gradient Boosting/XGBoost	82–90%, R^2 0.80–0.90	Moderate to high
Artificial Neural Networks (ANN)	80–88%	High
LSTM/CNN–LSTM	85–92%, with notable error reduction	High

Time Complexity of Similarity Measures after Normalization

$$\textbf{Time complexity} = O(n \times d)$$

Here, n = Samples & d = features (e.g. values of air pollutants like PM2.5, PM10, etc.) (Fig. 6).

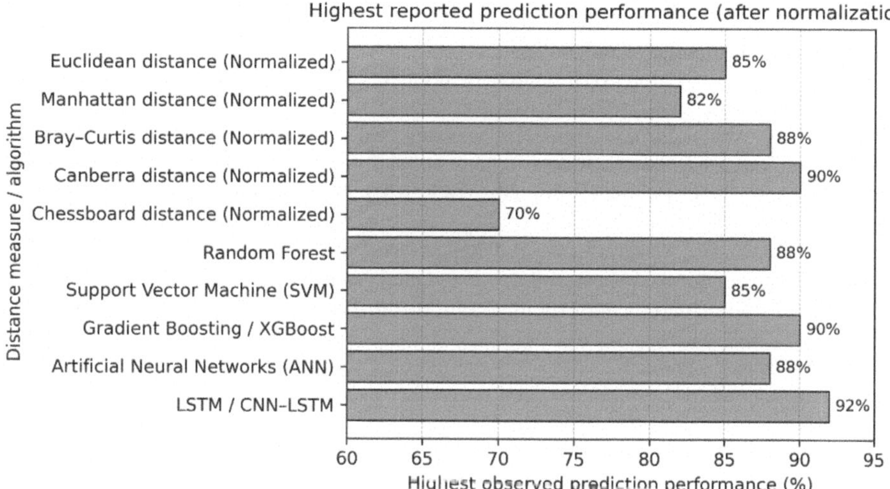

Fig. 6. Highest observed prediction performance for normalized similarity measures and machine learning techniques.

6 Research Gaps and Future Directions

Despite significant progress, several challenges remain in predicting respiratory and cardiovascular health risks due to air pollution. Many existing studies focus primarily on pollution prediction rather than direct health outcomes, limiting their utility in clinical and public health contexts. Additionally, there is a lack of standardized datasets linking high-resolution air quality data with long-term health records, particularly in low- and middle-income regions.

Future research is expected to move toward integrated frameworks that combine similarity measures, machine learning, and epidemiological knowledge to produce accurate, interpretable, and scalable health risk prediction models. Incorporating demographic vulnerability, mobility patterns, and multi-source sensor data can further enhance prediction accuracy and equity. Such models have the potential to support early warning systems, personalized health advisories, and data-driven policy interventions aimed at reducing the burden of air pollution-related diseases [29].

Acknowledgments. The authors extend their heartfelt gratitude to their home institutions for supporting and encouraging their research efforts.

Disclosure of Interests. The authors have no competing interests to declare that are relevant to the content of this article.

References

1. World Health Organization (WHO) Air pollution data portal Global Health Observatory (2025). https://www.who.int/data/gho/data/themes/air-pollution

2. World Health Organization (WHO) Global Air Quality Guidelines (2021). https://www.who.int/publications/i/item/9789240034228
3. World Health Organization (WHO) Global Air Quality Guidelines (AQG) (2021). https://tinyurl.com/3aaxrhz9
4. Farhadi, Z., Abulghasem Gorgi, H., Shabaninejad, H., et al.: Association between $PM_{2.5}$ and risk of hospitalization for myocardial infarction: a systematic review and a meta-analysis. BMC Public Health **20**, 314 (2020). https://doi.org/10.1186/s12889-020-8262-3
5. Yang, Y., Pei, Y., Gu, Y., Zhu, J., Yu, P., Chen, X.: Association between short-term exposure to ambient air pollution and heart failure: an updated systematic review and meta-analysis of more than 7 million participants. Front. Public Health **10**, 948765 (2023). https://doi.org/10.3389/fpubh.2022.948765
6. Zhou, W., Wen, Z., Peng, W., et al.: Association of ambient particulate matter with hospital admissions, length of hospital stay, and hospital costs due to cardiovascular disease: time-series analysis based on data from the Shanghai Medical Insurance System from 2016 to 2019. Environ. Sci. Eur. **35**, 46 (2023). https://doi.org/10.1186/s12302-023-00754-z
7. Sun, Y., et al.: Short term exposure to low level ambient fine particulate matter and natural cause, cardiovascular, and respiratory morbidity among US adults with health insurance: case time series study. BMJ **384**, e076322 (2024). https://doi.org/10.1136/bmj-2023-076322
8. Huangfu, P., Atkinson, R.: Long-term exposure to NO2 and O3 and all-cause and respiratory mortality: a systematic review and meta-analysis. Environ. Int. **144** (2020). https://doi.org/10.1016/j.envint.2020.105998
9. Yee, J., Cho, Y.A., Yoo, H.J., et al.: Short-term exposure to air pollution and hospital admission for pneumonia: a systematic review and meta-analysis. Environ. Health **20**, 6 (2021). https://doi.org/10.1186/s12940-020-00687-7
10. Lee, W., Lim, Y.H., Ha, E., et al.: Forecasting of non-accidental, cardiovascular, and respiratory mortality with environmental exposures adopting machine learning approaches. Environ. Sci. Pollut. Res. **29**, 88318–88329 (2022). https://doi.org/10.1007/s11356-022-21768-9
11. Shifaz, A., Pelletier, C., Petitjean, F., et al.: Elastic similarity and distance measures for multivariate time series. Knowl. Inf. Syst. **65**, 2665–2698 (2023). https://doi.org/10.1007/s10115-023-01835-4
12. Bhatt, N., et al.: Classification of Multi-date, Tempo-Spectral data using NDVI values. IJERA **3**(3), 174–179 (2013). www.ijera.com. ISSN 2248-9622
13. Kaggle Air Quality and Health Impact Dataset (2024). https://www.kaggle.com/datasets/rabieelkharoua/air-quality-and-health-impact-dataset
14. Pope, C.A., Dockery, D.W.: Health effects of fine particulate air pollution: lines that connect. J. Air Waste Manage. Assoc. (2006). https://pubmed.ncbi.nlm.nih.gov/16805397/
15. Balakrishnan, K., et al.: Exposure to household air pollution and risk of cardiovascular disease: findings from South Asia. Global Heart (2020)
16. UNICEF: Danger in the air: How air pollution can affect brain development in young children. https://www.unicef.org/sites/default/files/press-releases/glo-media-Danger_in_the_Air.pdf
17. Simoni, M., et al.: Adverse effects of outdoor pollution in the elderly. J. Thoracic Disease (2015). https://jtd.amegroups.org/article/view/3771/html
18. Schraufnagel, D.E., et al.: Air Pollution and Noncommunicable Diseases: Part 2: Air Pollution and Organ Systems. Chest (2019). https://pubmed.ncbi.nlm.nih.gov/30419237/
19. World Bank: The Cost of Air Pollution: Strengthening the Economic Case for Action. https://documents.worldbank.org/en/publication/documents-reports/documentdetail/781521473177013155
20. World Health Organization: WHO global air quality guidelines: particulate matter ($PM_{2.5}$ and PM_{10}), ozone, nitrogen dioxide, sulfur dioxide and carbon monoxide. World Health Organization, Geneva (2021)

21. Ballesteros Peinado, L., Guarda, T., Herrera-Vidal, G., Minnaard, C., Coronado-Hernández, J.R.: Statistical and machine learning models for air quality: a systematic review of methods and challenges. Algorithms **18**(12), 783 (2025). https://doi.org/10.3390/a18120783
22. Lee, W., Lim, Y.-H., Ha, E., Kim, Y., Lee, W.K.: Forecasting of non-accidental, cardiovascular, and respiratory mortality with environmental exposures adopting machine learning approaches. Environ. Sci. Pollut. Res. **29**, 88318–88329 (2022). https://doi.org/10.1007/s11356-022-21768-9
23. Psistaki, K., Richardson, D., Achilleos, S., Roantree, M., Paschalidou, A.K.: Assessing the impact of climatic factors and air pollutants on cardiovascular mortality in the Eastern Mediterranean using machine learning models. Atmosphere **16**(3), 325 (2025). https://doi.org/10.3390/atmos16030325
24. Madan, T., et al.: Hybrid deep learning model for air quality prediction and its impact on healthcare. Sci. Rep. **16**, 36564 (2026). https://doi.org/10.1038/s41598-026-36564-5
25. Koçak, E.: Comprehensive evaluation of machine learning models for real-world air quality prediction and health risk assessment by AirQ+. Earth Sci. Inform. **18**, 447 (2025). https://doi.org/10.1007/s12145-025-01941-7
26. Gryech, I., Asaad, C., Ghogho, M., Kobbane, A.: Applications of machine learning and IoT for outdoor air pollution monitoring and prediction: a systematic literature review. arXiv preprint arXiv:2401.01788 (2024)
27. Rajesh, M., Ganesh Babu, R., Moorthy, U., Easwaramoorthy, S.V.: Machine learning-driven framework for real-time air quality assessment and predictive environmental health risk mapping. Sci. Rep. **15**, 14214 (2025). https://doi.org/10.1038/s41598-025-14214-6
28. Houdou, A., et al.: Interpretable machine learning approaches for forecasting and predicting air pollution: a systematic review. Aerosol Air Qual. Res. **24**, 230151 (2024). https://doi.org/10.4209/aaqr.230151
29. Agbehadji, I.E., Obagbuwa, I.C.: Systematic review of machine learning and deep learning techniques for spatiotemporal air quality prediction. Atmosphere **15**(11), 1352 (2024). https://doi.org/10.3390/atmos15111352

PhD Forum 2025 - Development of Interpretable and Efficient Visual Question Answering Models

Souvik Chowdhury(✉) and Badal Soni

CSE Department, National Institute of Technology, Silchar, Assam, India
souvikcho@gmail.com

Abstract. Visual Question Answering (VQA) fuses vision and language to answer questions about images. Despite advances through deep learning, current models struggle with dataset bias, limited generalization, and high computational costs. We propose a multi-faceted solution to enhance VQA performance and efficiency. Our method reduces training and inference overhead via mixed-precision learning and question-type segregation. We design a hybrid attention architecture that integrates local-global visual features using co-attention, encoder-decoder, and spatial-channel mechanisms. To overcome language priors and reasoning limitations, we reframe VQA as a generative task using retrieval-augmented multimodal LLMs and prompt engineering. Our approach improves visual grounding, generalization, and scalability, contributing to robust and explainable VQA systems for real-world deployment. Our models achieve state-of-the-art performance on VQA v2.0, VQA-CP v2, GQA, and TDIUC benchmarks, showing up to 58% training speed-up and robust bias mitigation.

Keywords: Visual Question Answering · Multimodal Learning · Mixed Precision Training · Attention Mechanisms · Compositional Reasoning · Language Prior · Large Language Models

1 Introduction

Visual Question Answering (VQA) is a multimodal task that requires understanding both images and natural language to generate accurate answers. Unlike tasks such as image captioning or object detection, VQA demands fine-grained reasoning—recognizing objects, spatial relations, and contextual cues. While CNNs and transformer-based models have advanced VQA performance, current methods face key challenges: (1) over-reliance on textual cues (language bias), (2) limited ability in compositional reasoning, and (3) high computational demands. VQA holds promise for real-world applications such as accessibility, surveillance, healthcare, and education. However, existing models often fail to adequately ground answers in visual content and struggle with multi-step or relational queries. Moreover, their efficiency remains a bottleneck in practical deployment. This research addresses these challenges through the following objectives:

© The Author(s), under exclusive license to Springer Nature Switzerland AG 2026
A. Kannan et al. (Eds.): ADCOM 2025, CCIS 2947, pp. 228–238, 2026.
https://doi.org/10.1007/978-3-032-26269-1_16

1. Develop a generalized VQA framework with improved execution efficiency.
2. Propose an attention-based architecture that enhances semantic alignment between visual and textual features.
3. Mitigate language bias by reinforcing visual grounding in answer generation.
4. Enable compositional reasoning to handle complex, multi-object queries.

1.1 Unified Modular VQA Framework

Although each component of the proposed system addresses a distinct limitation of contemporary Visual Question Answering (VQA) models, the overall design follows a unified and modular architectural philosophy. Rather than operating as independent model variants, the proposed techniques function as coordinated layers within a single computational pipeline.
The framework is organized into four functional stages:

1. **Computational Efficiency Layer** – Mixed-precision optimization and question-type segregation reduce training overhead and enable scalable model specialization.
2. **Visual Grounding Layer** – A hybrid attention architecture integrates global and region-level representations to improve alignment between visual evidence and linguistic queries.
3. **Reasoning and Generative Inference Layer** – Retrieval-augmented prompting and contextual information synthesis support compositional reasoning and mitigate language priors.
4. **Dataset and Evaluation Layer** – A bias-aware dataset and semantic similarity-based evaluation framework enable robust assessment of open-ended answer generation.

These layers operate sequentially and share intermediate representations, forming a coherent end-to-end system that jointly optimizes efficiency, interpretability, and reasoning capability. This architectural unification distinguishes the proposed approach from prior work that addresses these challenges in isolation.

1.2 Contributions

This work makes the following technical contributions to the development of interpretable and efficient Visual Question Answering systems:

1. We propose a unified modular VQA architecture that jointly addresses computational efficiency, visual grounding, and compositional reasoning within a single end-to-end framework.
2. We introduce a computational optimization strategy combining automatic mixed-precision training with question-type-aware model specialization, resulting in substantial reductions in training and inference overhead while preserving predictive performance.

3. We develop a hybrid spatial–channel co-attention architecture that integrates global scene context with region-level object representations, improving visual grounding and reducing susceptibility to language bias.
4. We formulate VQA as a structured generative reasoning process using retrieval-augmented contextual prompting, enabling improved handling of compositional and multi-entity queries.
5. We propose a semantic similarity-based evaluation metric that measures answer correctness in embedding space, providing a more faithful assessment of generative model outputs than strict lexical matching.
6. We construct a bias-aware VQA dataset annotated for language prior sensitivity and compositional reasoning complexity, supporting more rigorous evaluation of model generalization.

Section 2 reviews related work. Section 3 discusses efficiency techniques. Section 4 details the hybrid attention architecture. Sections 5 and 6 address bias and reasoning using generative models. Section 7 concludes with key insights and future directions. Appendix A and B provides the important publications as an outcome of this research and detailed results respectively.

2 Literature Survey

VQA research has progressed from early CNN-LSTM fusion models to advanced architectures targeting reasoning and bias mitigation. Attention mechanisms—top-down, bottom-up, and co-attention—improved localization and modality alignment [2]. Multimodal bilinear pooling (MFB, MFH) and learnable fusion strategies enhanced image-text interactions [9]. Fine-grained attention, including spatial, channel, and temporal variants, enabled more precise reasoning [4]. Transformer-based models and dense attention have introduced global context and causal reasoning capabilities [10]. Compositional reasoning gained focus with datasets like CLEVR, pushing multi-entity relational understanding [14]. To improve inference, multi-hop reasoning and external knowledge sources have been explored [7]. Yet, bias persists—models often rely on language priors. Debiasing techniques, including ensemble methods, counterfactual training, and justification generation, have emerged to promote visually grounded reasoning [6].

3 Improving VQA Model's Computation Time

We propose two complementary methods—AMPVQA and QSFVQA to improve the efficiency and scalability of Visual Question Answering (VQA) models. AMPVQA is a wrapper-based framework that accelerates training by integrating automatic mixed precision (AMP). It applies FP16 precision for gradient computation and backpropagation, FP32 for weight updates and loss calculation, and employs dynamic loss scaling to prevent underflow issues. We evaluated our models on VQA v2.0 and CLEVR datasets, AMPVQA significantly reduces training time and memory consumption while preserving accuracy. QSFVQA further enhances performance and efficiency by segregating questions by type. It automatically classifies question categories (e.g., *Color*, *Sport*, *Gender*) using TF-IDF features and a Naive Bayes classifier, enabling a modular and type-specific training pipeline. AMPVQA and QSFVQA together present a robust approach for enhancing the training and inference efficiency of VQA systems. AMPVQA offers a 33% reduction in training time, while QSFVQA further improves execution time by 26%, achieving a total speed-up of 58% with marginal trade-offs in accuracy (Fig. 1).

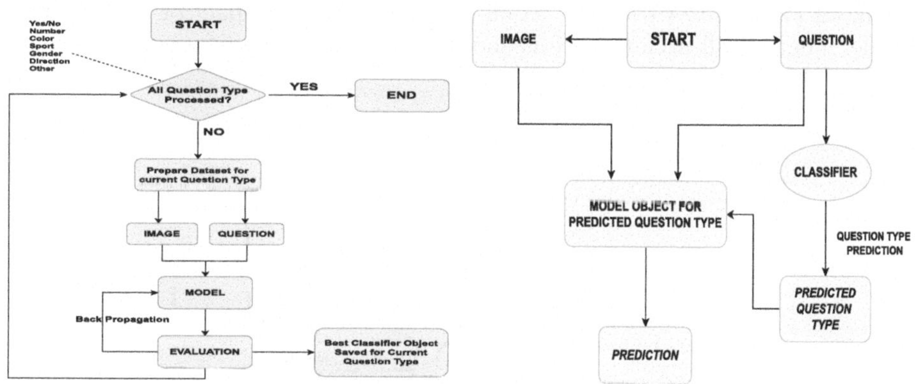

Fig. 1. Training and Inference Pipeline in QSFVQA

3.1 System-Level Integration

Each component of the proposed framework contributes to a distinct stage of the VQA pipeline. Their functional roles and interactions are summarized as follows.

The efficiency layer operates during training and inference to reduce computational cost through precision optimization and question-type specialization. The visual grounding layer processes image and question representations to produce aligned multimodal embeddings. These grounded representations are then supplied to the reasoning layer, where contextual information retrieval and structured prompting support answer generation. Finally, the dataset and evaluation layer provides bias-sensitive supervision and semantic assessment of model outputs.

This staged design ensures that improvements in efficiency, grounding, and reasoning are not independent enhancements but mutually reinforcing processes within a unified architecture.

4 Attentive VQA: Precision Through Focus

Attention mechanisms in Visual Question Answering (VQA) are typically categorized into spatial, channel, or temporal types, with many models using only limited combinations [11]. We propose ESCNet, a hybrid attention-based architecture that integrates global and region-level features through spatial and channel attention. Its core Attention Ensemble Module fuses coarse-grained scene context from ResNet with fine-grained object features from Faster R-CNN, enhancing visual grounding and addressing spatial ambiguity. ESCNet's dual and triplet attention modules capture interactions across height, width, and channel dimensions, improving feature expressiveness. A hierarchical co-attention network aligns visual features with DistilBERT-derived question embeddings, enabling stronger cross-modal reasoning. ESCNet achieves +3.8% on VQA-CP v2 and +2.7% on GQA over the baseline, validating the impact of the ensemble module. The overall architecture is shown in Fig. 2, with performance comparisons in Table 2.

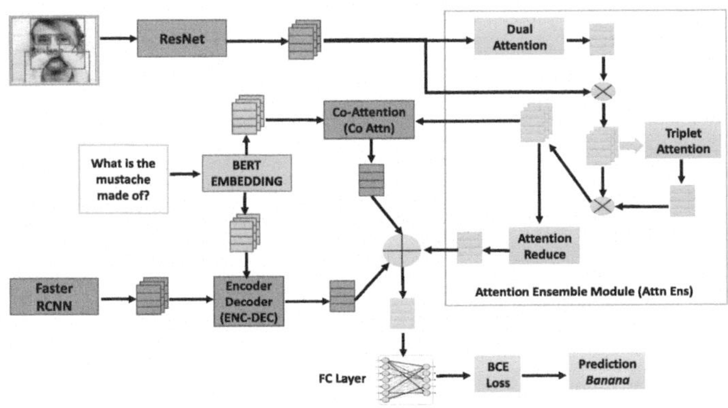

Fig. 2. Proposed ESCNet Model Architecture

5 Mitigating Language Prior

We address the language prior problem in Visual Question Answering (VQA), where biased answer distributions—especially for ambiguous or multi-modal queries—undermine generalization. To mitigate this, we adapt our proposed ESC-Net model to emphasize visually grounded reasoning by focusing attention on image regions relevant to the question. Evaluations on VQA-CP v2 and VQA v2 demonstrate the effectiveness of this approach. ESC-Net achieves 78.56% on VQA-CP v2 and 83.57% on VQA v2, outperforming baselines like BUTD, RUBi, LMH, and CSS, particularly in bias-prone categories such as *Number* and *Other* (Table 2). Attention heatmaps (Table 3) reveal that ESC-Net accurately highlights semantically relevant regions, reducing reliance on linguistic priors. For example, it correctly identifies grey as the color of a cloudy sky where biased models typically predict blue. These results confirm ESC-Net's robustness against bias through enhanced visual grounding. Future directions include refining attention for multi-object scenes and integrating contrastive supervision to further suppress priors (Figs. 4 and 5).

6 Addressing Compositional Reasoning

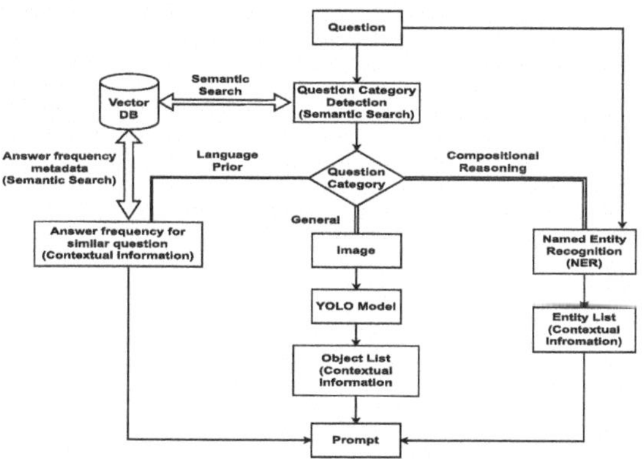

Fig. 3. Intelligent Heuristic Prompt Module

Despite progress in Visual Question Answering (VQA), addressing both language priors and compositional reasoning in a unified framework remains a challenge. We introduce GenVQA, a structured four-stage pipeline comprising: (1) Question Category Identification, (2) Contextual Information Generation, (3) Prompt Generation, and (4) Prediction. To improve evaluation, we propose a semantic similarity-based accuracy metric that goes beyond exact-match limitations. We also present a new benchmark dataset with flexible preparation and LLM-assisted annotation moderation to ensure quality. Our R-VQA model enhances text-grounded visual understanding and supports open-ended answer generation. Figure 3 illustrates the GenVQA architecture. Traditional exact-match metrics often misjudge semantically correct responses from Large Language Models (LLMs) that don't precisely match the ground truth—for example, treating "It is blue" as incorrect for the answer "blue." To address this, we propose a semantic accuracy metric that computes cosine similarity between the embedding vectors of predicted and reference answers. A prediction is considered correct if the similarity exceeds a threshold of 0.9. The redefined metric is expressed as:

$$\text{Accuracy} = \min\left[\frac{\text{count}\left(\cos\left(\text{embed}(\text{predicted\_ans}), \text{embed}(\text{ground\_truth})\right) > 0.9\right)}{3}, 1\right] \quad (1)$$

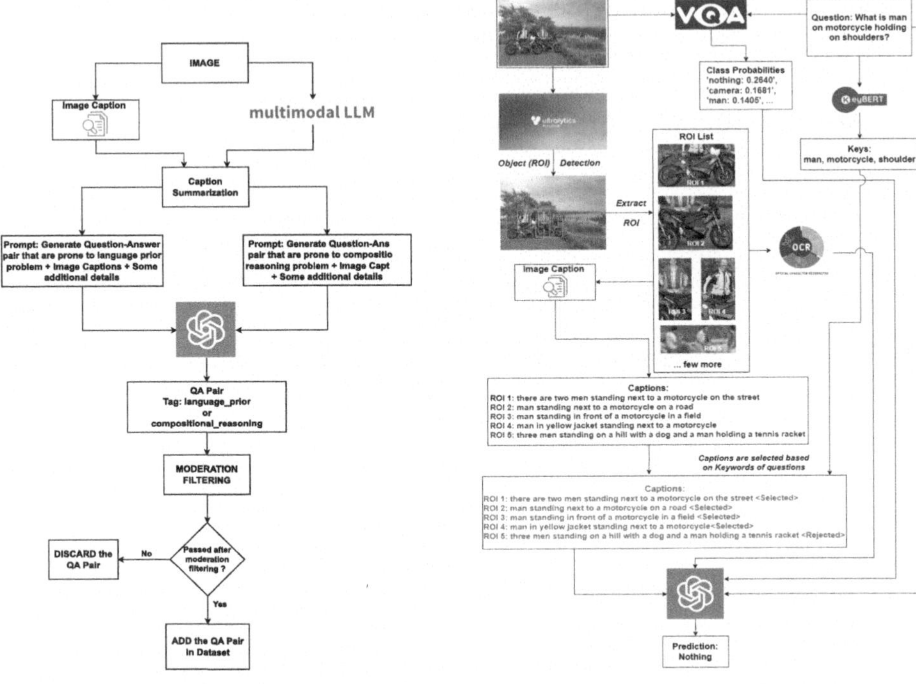

Fig. 4. Dataset Preparation Workflow

Fig. 5. R-VQA Model Architecture

7 Conclusion and Future Scope

This research advances VQA by addressing four core challenges: overreliance on text, language bias, limited compositional reasoning, and inefficiency. We introduce AMPVQA and QSFVQA for faster, modular training; ESC-Net for enhanced spatial-channel attention and visual grounding; and heuristic prompting for zero-shot generalization in large multimodal models. While gains are observed, multi-object ambiguity remains a challenge. Our models, metrics, and dataset innovations set new benchmarks. Future directions include video-based temporal reasoning, hierarchical fusion, multilingual datasets, edge deployment, and dialog-enabled multimodal systems for broader real-world impact.

Important Publications

1. Souvik Chowdhury and Badal Soni, **"QSFVQA: A Time Efficient, Scalable and Optimized VQA"**, *Arabian Journal for Science and Engineering*, Springer (2023),
DOI: https://doi.org/10.1007/s13369-023-07661-8, [SCI].
2. Souvik Chowdhury, Badal Soni, and Doli Phukan, **"Incorporation of Question Segregation Procedures in Visual Question-Answering Models"**, *International Journal of Computing Science and Mathematics*, Inderscience Publishers (IEL) (2024),
DOI: https://doi.org/10.1504/IJCSM.2024.140859, [SCOPUS].
3. Souvik Chowdhury and Badal Soni, **"Beyond Words: ESC-Net Revolutionizes VQA by Elevating Visual Features and Defying Language Priors"**, *Computational Intelligence*, Willey (2024), Volume 40, Issue 6, DOI: https://doi.org/10.1111/coin.70010 [SCI].
4. Souvik Chowdhury and Badal Soni, **"A Robust Visual Question Answering Model"**, *Knowledge-Based Systems*, Elsevier (2024), Volume 309,
DOI: https://doi.org/10.1016/j.knosys.2024.112827 [SCI].
5. Souvik Chowdhury and Badal Soni, **"ENVQA: Improving Visual Question Answering model by enriching the visual feature"**, *Engineering Applications of Artificial Intelligence*, Elsevier (2024), Volume 142, DOI: https://doi.org/10.1016/j.engappai.2024.109948 [SCI].
6. Souvik Chowdhury and Badal Soni, **"Handling Language Prior and Compositional Reasoning Issues in Visual Question Answering System"**, *Neurocomputing*, Elsevier (2025), Volume 635, DOI: https://doi.org/10.1016/j.neucom.2025.129906 [SCI].
7. Badal Soni and Souvik Chowdhury, **"A System for an Efficient and Improved Visual Question Answering Framework"**, Patent number *DE 20 2023 100 081 U1*, Granted on 2023, *[PATENT]*.
8. Souvik Chowdhury, Badal Soni, November 27, 2024, "Robust VQA (RVQA) Dataset with Language Prior and Compositional Reasoning Labels", IEEE Dataport, DOI: https://dx.doi.org/10.21227/9cjm-dx19, [DATASET].

Final Result and Examples

Table 1

Table 1. Summary of Core Contributions

Module	Challenge Addressed	Key Technique	Impact
AMPVQA	High training time	Mixed-precision training (FP16 for gradients, FP32 for weights)	33% reduction in training time
QSFVQA	Scalability and redundancy in question types	Question-type classification and modular pipeline	+26% faster inference
ESC-Net	Weak visual grounding, language bias	Hybrid spatial–channel co-attention with ensemble alignment	+3–4% accuracy on VQA-CP and GQA
GenVQA	Language prior and compositional reasoning	Retrieval-augmented LLM prompts with heuristic context generation	Robust zero-shot generalization and bias mitigation
R-VQA	Lack of robust datasets	Curated dataset with semantic filtering and compositional tags	Benchmarked on 5 datasets with state-of-the-art results

Table 2. Final Performance Comparison (% Accuracy)

Model	VQA v2.0	VQA-CP	GQA	TDIUC	Visual7W
Flamingo [1]	82.1	–	–	–	–
CoCa [16]	82.3	–	–	–	–
ONE-PEACE [13]	82.6	–	–	–	–
BAN2_CTI [5]	70.04	61.27	–	87.0	–
Coarse-to-Fine [12]	–	–	72.14	–	71.9
PEVL+ [15]	–	–	77.0	–	–
CMN [8]	–	–	–	–	72.53
CTI [5]	–	–	–	–	72.3
LLM Prompt [10]	–	–	72.66	–	–
MUREL [3]	–	–	–	88.2	–
ESC-Net (Ours)	**83.57**	**78.56**	**73.04**	**87.63**	–
GenVQA (Ours)	**82.37**	**77.51**	**77.49**	**88.90**	**79.68**
R-VQA (Ours)	**84.04**	**79.08**	**79.65**	**89.7**	**81.17**

Table 3. Model-wise Example Outputs

Model Name	**Example Walkthrough**
ESCNet	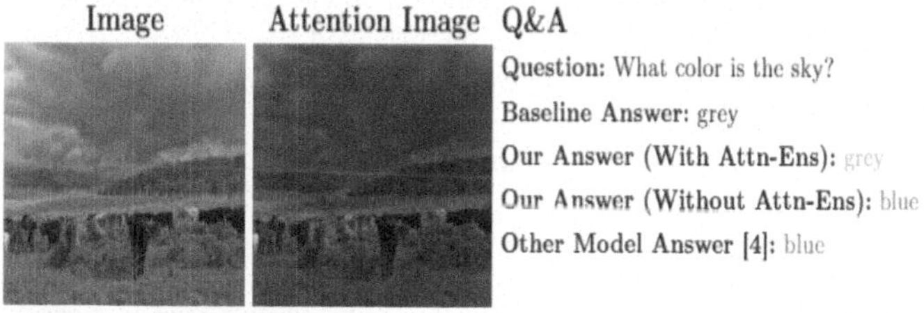 Image Attention Image Q&A **Question:** What color is the sky? **Baseline Answer:** grey **Our Answer (With Attn-Ens):** grey **Our Answer (Without Attn-Ens):** blue **Other Model Answer [4]:** blue
GenVQA	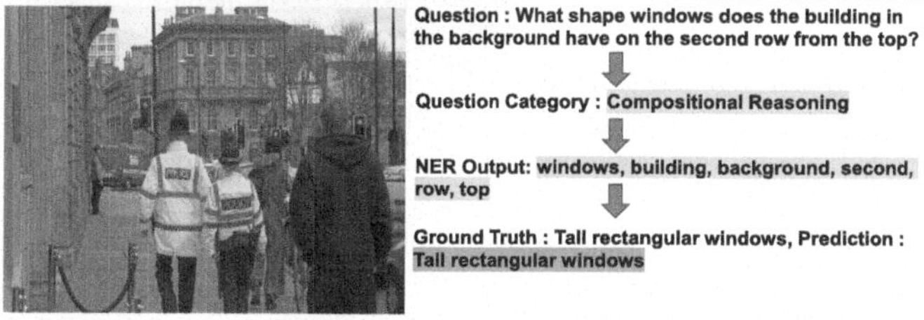 **Question : What shape windows does the building in the background have on the second row from the top?** **Question Category : Compositional Reasoning** **NER Output: windows, building, background, second, row, top** **Ground Truth : Tall rectangular windows, Prediction : Tall rectangular windows**
RVQA	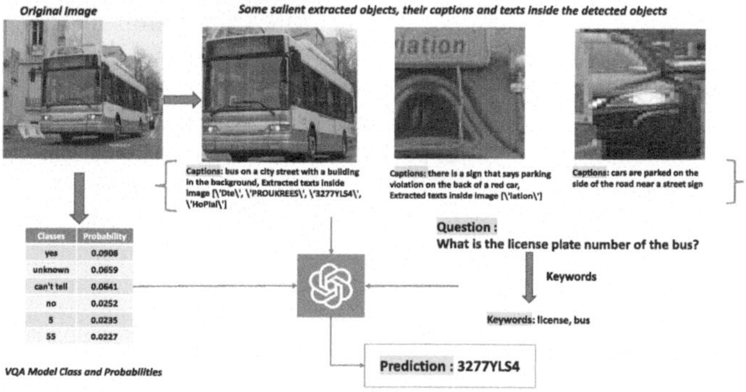 *Original Image* *Some salient extracted objects, their captions and texts inside the detected objects* Captions: bus on a city street with a building in the background, Extracted texts inside image [\'Dte\', \'PROUKREES\', \'3277YLS4\', \'HoPlal\'] Captions: there is a sign that says parking violation on the back of a red car, Extracted texts inside image [\'lation\'] Captions: cars are parked on the side of the road near a street sign Classes / Probability: yes 0.0906; unknown 0.0659; can't tell 0.0641; no 0.0252; 5 0.0235; 55 0.0227 *VQA Model Class and Probabilities* Question : What is the license plate number of the bus? Keywords Keywords: license, bus Prediction : 3277YLS4

References

1. Alayrac, J.B., et al.: Flamingo: a visual language model for few-shot learning. Adv. Neural. Inf. Process. Syst. **35**, 23716–23736 (2022)
2. Anderson, P., He, X., Buehler, C., Teney, D., Johnson, M., Gould, S., Zhang, L.: Bottom-up and top-down attention for image captioning and visual question answering. In: Proceedings of the IEEE Conference on Computer Vision and Pattern Recognition, pp. 6077–6086 (2018)
3. Cadene, R., Ben-Younes, H., Cord, M., Thome, N.: Murel: multimodal relational reasoning for visual question answering. In: Proceedings of the IEEE/CVF Conference on Computer Vision and Pattern Recognition, pp. 1989–1998 (2019)
4. Chen, C., Han, D., Shen, X.: Clvin: complete language-vision interaction network for visual question answering. Knowl.-Based Syst. **275**, 110706 (2023)
5. Do, T., Do, T.T., Tran, H., Tjiputra, E., Tran, Q.D.: Compact trilinear interaction for visual question answering. In: Proceedings of the IEEE/CVF International Conference on Computer Vision, pp. 392–401 (2019)
6. Gao, D., Wang, R., Shan, S., Chen, X.: Cric: a vqa dataset for compositional reasoning on vision and commonsense. IEEE Trans. Pattern Anal. Mach. Intell. **45**(5), 5561–5578 (2022)
7. Guo, D., Tao, D.: Learning compositional representation for few-shot visual question answering. arXiv preprint arXiv:2102.10575 (2021)
8. Hu, R., Rohrbach, M., Andreas, J., Darrell, T., Saenko, K.: Modeling relationships in referential expressions with compositional modular networks. In: Proceedings of the IEEE Conference on Computer Vision and Pattern Recognition, pp. 1115–1124 (2017)
9. Li, X., et al: Learnable aggregating net with diversity learning for video question answering. In: Proceedings of the 27th ACM International Conference on Multimedia, pp. 1166–1174 (2019)
10. Liu, J., Fang, C., Li, L., Li, B., Hu, D., Ma, C.: Prompting large language models with fine-grained visual relations from scene graph for visual question answering. In: ICASSP 2024-2024 IEEE International Conference on Acoustics, Speech and Signal Processing (ICASSP), pp. 8125–8129. IEEE (2024)
11. Mishra, A., Anand, A., Guha, P.: Dual attention and question categorization-based visual question answering. IEEE Trans. Artif. Intell. **4**(1), 81–91 (2022)
12. Nguyen, B.X., Do, T., Tran, H., Tjiputra, E., Tran, Q.D., Nguyen, A.: Coarse-to-fine reasoning for visual question answering. In: Proceedings of the IEEE/CVF Conference on Computer Vision and Pattern Recognition, pp. 4558–4566 (2022)
13. Wang, P., et al.: One-peace: exploring one general representation model toward unlimited modalities. arXiv preprint arXiv:2305.11172 (2023)
14. Yan, X., et al.: Comprehensive visual question answering on point clouds through compositional scene manipulation. IEEE Trans. Vis. Comput. Graph. (2023)
15. Yao, Y., Chen, Q., Zhang, A., Ji, W., Liu, Z., Chua, T.S., Sun, M.: Pevl: position-enhanced pre-training and prompt tuning for vision-language models. arXiv preprint arXiv:2205.11169 (2022)
16. Yu, J., Wang, Z., Vasudevan, V., Yeung, L., Seyedhosseini, M., Wu, Y.: Coca: Contrastive captioners are image-text foundation models. arXiv preprint arXiv:2205.01917 (2022)

Author Index

B
Bhatt, Neha 215
Bora, Anubrat 160, 173

C
Chandan, Gudimetla Lakshmidhara 23
Chaturvedi, Ashvini 115
Chitra, A. 36, 201
Chowdhury, Souvik 228

D
D'Souza, Meenakshi 3
Deepak, Gerard 147, 160, 173
Dharshini, S. Priya 36
Dhiman, Nitin Kumar 51

G
Gayathri, K. 36, 201

H
Hati, Avik 183

I
Impana, K. P. 130

J
Joseph, Asha 215

K
Kaliyari, Dushyant 66
Kalra, Aniket 51
Karampuri, Ramsha 87
Khemani, Deepak 51
Kulkarni, Prakash S. 87

L
Lenkalapelly, Raju 87
Lokhande, Makarand M. 87

M
Mahima 3
Mishra, Sachin 99

N
Naga Siva Prasad, Bodempudi 115
Nusrath, Khadeeja 66

P
Palanisamy, P. 183
Prajwal, C. A. 130

R
Ramnathan, Malmathanraj 183
Rashmi, H. 115
Roy, Anindita 87

S
Sahu, Akash 66
Salman, S. A. Mohammed 147
Santhanavijayan, A. 147
Sathish Kumar, A. 183
Sawarkar, Pravin D. 87
Shukla, Samiksha 160
Singh, Abhinav Kumar 23
Soni, Badal 228

T
Trishank, K. Bala 23

V
Vijayakumar, Vaidehi 23

© The Editor(s) (if applicable) and The Author(s), under exclusive license
to Springer Nature Switzerland AG 2026

A. Kannan et al. (Eds.): ADCOM 2025, CCIS 2947, p. 239, 2026.
https://doi.org/10.1007/978-3-032-26269-1

The manufacturer's authorised representative in the EU is Springer Nature Customer Service Centre GmbH, Europaplatz 3, 69115 Heidelberg, Germany. If you have any concerns regarding our products, please contact ProductSafety@springernature.com

Printed and bound by CPI Group (UK) Ltd, Croydon, CR0 4YY

15/07/2026

02167586-0001